Reframing the Ethnographic Museum

Reframing the Ethnographic Museum

Histories, politics and futures

Edited by Michael Rowlands, Nick Stanley
and Graeme Were

UCLPRESS

First published in 2025 by
UCL Press
University College London
Gower Street
London WC1E 6BT

Available to download free: www.uclpress.co.uk

A CIP catalogue record for this book is available from The British Library.

ISBN: 978-1-80008-588-6 (Hbk)
ISBN: 978-1-80008-587-9 (Pbk)
ISBN: 978-1-80008-586-2 (PDF)
ISBN: 978-1-80008-589-3 (epub)
DOI: https://doi.org/10.14324/111.9781800085862

Contents

Part III: Future directions

List of figures

List of contributors

Michael Rowlands is an emeritus professor of Material Culture, Department of Anthropology, University College London. His research interests include the theorisation and conceptualisation of cultural heritage and material culture. He has conducted fieldwork research on the histories and development of postcolonial museums in West Africa, the role of cultural heritage in post-conflict Liberia and is currently doing fieldwork on private museum initiatives in Sichuan and Yunnan provinces in China, funded by the Leverhulme Foundation. He collaborates with anthropologists in the National University of Taiwan on the revitalisation of indigenous cultural knowledge. He is also a visiting professor at the University of Cape Town where he teaches the role of cultural heritage and museums in postcolonial Africa. He is a visiting professor to the University of Gothenburg, Sweden, consulting on the development of a cultural heritage initiative and a Graduate School programme in cultural heritage.

Nick Stanley is an emeritus professor at Birmingham City University and currently honorary research fellow at the Department of Africa, Oceania and the Americas at the British Museum.

Graeme Were is head of the College of Law, Anthropology and Politics and a professor of Anthropology at SOAS, University of London. He has a regional specialism in Vietnam and Papua New Guinea, where he has conducted long-term ethnographic research on museums, material culture and heritage. His published work includes *Museums, Collections and Social Repair in Vietnam* (Routledge, 2022) and *How Materials Matter: Design, innovation and materiality in the Pacific* (Berghahn, 2019).

Ben Burt is an anthropologist who has worked for the British Museum since 1974 as an educator and curator. He has researched with the Kwara'ae people of Malaita in the Solomon Islands since 1979 and published a number of books and articles on Malaitan history, culture, artefacts and oral literature, as well as a textbook on world art and a memoir about the British Museum's Museum of Mankind. At present he works part time for the Oceania section of the British Museum Department of Africa, Oceania and the Americas.

Alison Clark is Senior Curator Oceania at National Museums Scotland. She has 15 years' experience working in the museum sector including the British Museum and the Museum of Archaeology and Anthropology, University of Cambridge. Since 2009 she has worked collaboratively with indigenous communities in Oceania to document the contemporary legacies of colonial collecting in the region. Her research is focused on Australia and the Pacific Islands where she has a particular interest in environmental change, biocultural collections and the history of collections.

Shona Coyne is head of Shared Histories at the Western Australian Museum, Perth. Shona is a Menang/Nyungar woman with cultural connections to Yamatji Country in Western Australia and the Scottish Highlands. She is also a senior curator and manager of Repatriation and Community Engagement at the National Museum of Australia. Her research interests centre around the intersections between Indigenous and non-Indigenous colonial histories. Her recent museum projects include the award-winning exhibitions *Endeavour Voyage: The Untold Stories of Cook and the First Australians* and *Belonging: Stories from Far North Queensland* at the National Museum of Australia. She also curated the exhibition *Yurlmun: Mokare Mia Boodja* through the West Australian Museum. Recent publications include 'New postings? The Swan River Colony' with Tiffany Shellam in *Ancestors, Artefacts, Empire: Indigenous Australia in British and Irish museums* (British Museum Press, 2021).

Ferdinand de Jong is a senior lecturer in anthropology at the University of East Anglia, Norwich, with a doctorate from the University of Amsterdam. He is a cultural anthropologist whose research interests include museums, memory, cultural heritage, decolonisation, religion, and repair. De Jong's latest monograph is *Decolonizing Heritage: Time to repair in Senegal*, published by Cambridge University Press (2022). His most recent edited volume (with José Mapril) addresses religious heritage in a post-secular world, *The Future of Religious Heritage: Entangled temporalities of the sacred and the secular* (Routledge, 2023).

Taku Iida is an ecological and cultural anthropologist at the National Museum of Ethnology, Japan (NME). Under NME's Info-Forum

Museum Project, he tries to establish long-lasting communication with communities from which museum objects derived, and to promote global sharing of local value and knowledge. His recent interests also include the cognitive base of human creativity, visual media, cultural transmission, and history of Japanese anthropology. He is the editor of *Heritage Practices in Africa* (National Museum of Ethnology, 2022), *Living with Cultural Heritage* (in Japanese, Rinsen Book, 2017), and *Cultural Heritage in Human History* (in Japanese, Rinsen Book, 2017).

Chantal Knowles is a museum professional with over 25 years of experience working with cultural collections in museums in the United Kingdom, Australia and Aotearoa, New Zealand. She has developed permanent galleries at National Museums Scotland and the Queensland Museum. She is head of Human History at Tāmaki Paenga Hira Auckland War Memorial Museum and co-director of Te Aho Mutunga Kore, a textile and fibre knowledge exchange centre. The centre is kaupapa Māori-led and prioritises and funds Māori and Pacific artists, makers, practitioners and knowledge holders' access to museum collections. She is interested in the role of objects and collections in creating and sustaining narratives.

Arnaud Lismond-Mertes is the Secrétaire Général, Collectif Solidarité contre l'Exclusion, Bruxelles.

Sean Mallon is Senior Curator for Pacific Cultures at the Museum of New Zealand Te Papa Tongarewa, Wellington. He specialises in the social and cultural history of Pacific peoples in New Zealand. He is the author of *Tatau: A history of Sāmoan tattooing* (2017) with Sébastien Galliot, and is currently researching issues relating to the agency and activism of Pacific peoples in museums.

Luo Pan received her doctoral degree in 2011 from the Chinese University of Hong Kong, Hong Kong SAR, as an anthropology major. Since 2014 she has been an associate research fellow in the Research Division of the Chinese National Museum of Ethnology, Beijing.

Alistair Paterson is an archaeologist, Australian Research Council (ARC) Future Fellow and professor at the University of Western Australia School of Social Sciences. His research examines the historical archaeology of colonial coastal contact and settlement in Australia's Northwest and the Indian Ocean.

Tiffany Shellam is an associate professor, and lectures in History at Deakin University. She works collaboratively with Noongar people, historians, curators and archivists to critique archives and collections, unearthing hidden and alternative histories generated by encounters between First Nation people and explorers and colonists. Her book *Meeting the Waylo: Aboriginal encounters in the archipelago* (UWAP, 2019) won the Prime Minister's Prize for Australian History in 2020.

Endah Triastuti graduated from the University of Wollongong, New South Wales, Australia and received her PhD from the School of Social Science, Media and Communication, Faculty of Arts. She is now an associate professor at Universitas Indonesia, Jakarta. She is a senior research fellow at the Asia Research Center, Universitas Indonesia. Her core research interests include digital ethnography, gender and minority and media anthropology.

Boris Wastiau pursued arts and anthropology studies at the Université libre de Bruxelles, the University of Coimbra, and the University of East Anglia. Following that, he served as a curator at the Royal Museum for Central Africa in Belgium from 1996 to 2007. Later, Wastiau became the director of the Musée d'Ethnographie de Genève (MEG), where he spearheaded its expansion, restructuring and reorientation, leading to the MEG receiving the prestigious European Museum of the Year Award in 2017. Throughout his career, Wastiau has authored publications on African religions, shamanism, colonialism and the history of colonial museologies. Additionally, he taught African religions and African arts at the University of Geneva from 2009 to 2019. Since 1 April 2022 Wastiau has held the position of director at the Alimentarium Foundation in Vevey, Switzerland.

Part I
The historic legacy of ethnographic museums

1

The changing politics of ethnographic display: a review

Michael Rowlands, Nick Stanley and Graeme Were

Ethnographic museums have suffered a poor reputation amongst anthropologists and museum curators alike for as long as they have existed. Such museums and their displays have been scorned for their inevitable association with the racism endemic in imperial and colonial rule over subject peoples. Ethnographic museums were inexorably linked to imperialism (Kuper 2023, 37), while critics have been even more explicit. Hicks, for instance, speaks of the anthropological museum during the 1890s as 'a weapon, a method and a device for the ideology of white supremacy to legitimate, extend and naturalise the new extremes of violence within corporate colonialism' where 'archaeology and anthropology came to be tools in subjugation through the selection and display of material culture' (Hicks 2020, 15, 181). Similarly, Lidchi and Allan remark that museums have become synonymous with 'plunder, booty, loot and trophies' (2020, 54).

It has been argued that 'empire building and museum building went hand-in-hand' (Aldrich 2009, 138). This may be a claim that requires further thought, as these two events did not always occur synchronically. Explorers, engineers, administrators and others made their collections in the period of high imperialism, but their collections did not necessarily reach museums until after they had retired from the field, frequently only as posthumous bequests from their executors. There may, therefore, be a disjunction between the making of the collections and their appearance in the museum. At this latter point two further problems arose. The collector's initial objectives were seldom communicated to the curators who received them, and the sheer pressure of bequests from colonial settings meant that they lay dormant and uncatalogued. Curators risked being overwhelmed by their bounty and unable to respond adequately.

There arose an unstoppable collecting mania among curators who argued that 'collecting had to remain their top priority, and [that] it was better to postpone their cataloguing, and simply pack their artefacts away in boxes, rather than limit their acquisition' (Penny 2002, 183).

This disjuncture between the intentions of the collectors and the recipient curators resulted in the artefacts being 'defined, segmented, detached and carried away by ethnographers' (Kirschenblatt-Gimblet 1991, 387) and marks a major failure to integrate colonial artefacts and collections into any coherent narrative. The picture remains too opaque and fragmentary. This is not to suggest that imperial ambitions were forgotten in the displays that did appear. Indeed, the dioramas into which the artefacts were placed frequently underlined the division between imperial conquerors and 'primitive peoples' who, as Bennett puts it, 'dropped out of history altogether in order to occupy a twilight zone between nature and culture' (Bennett 1995, 77). The blend between natural history and anthropology became blurred in major displays of colonial life.

Since the end of the nineteenth century there have been many and various attempts to eradicate this stigma, but results have not really dislodged such initial perceptions. The problem starts from the very first exhibitions. The ethnographic displays at the Chicago World's Fair of 1893 gave rise to both the seriousness of Franz Boas's American Museum of Natural History as well as to the entertainment of P. T. Barnum's American Museum. Similarly, in Britain the colonial context within which ethnographic displays were set created much discomfort at the racist typologies employed (Altick 1978). This led to the widespread abandonment of ethnographic exhibitions from the middle of the twentieth century. A renaissance only became possible once a new philosophical system emerged – multiculturalism – which gave rise to such institutions as the Museum of Mankind (whose title in turn raised further dispute in terms of gender equality) (see Chapters 2 and 3) and to 'World Museums', both in Britain but elsewhere across the world (see Chapters 5 to 10). Another buffeting has occurred in the past decade as subaltern theory has percolated through into ethnographic display. This raises fresh issues for ethnographic exhibits that seek to shore up nationalist sentiment through calls to 'ethnic pluralism'. Perhaps now is a good time to consider how ethnographic display has survived and, perhaps as importantly, where and how it has failed. Is ethnographic display still viable today, and if so, under what circumstances and conditions?

It is now nearly a quarter of a century since Anthony Shelton's 'The future of museum ethnography', which questioned the future relevance of ethnographic museums. His condemnation was emphatic: 'It would perhaps be no exaggeration to see museum ethnography as methodologically bankrupt, practically or performatively outdated, and its public product as deceptive of the "reality" it claims to represent' (Shelton 1997, 45). Furthermore, it is nearly a decade since the end of the EU-funded RIME (Ethnography Museums and World Cultures) project and the subsequent Pitt Rivers Museum conference, and also since Clare Harris and Michael O'Hanlon's article 'The future of ethnographic museums'. Now the time seems ripe to question whether Shelton's claim that 'a line of new leaders [has appeared] whose work shows great promise in facing the far greater complexities of history, society and culture' (Shelton 1997, 45) can be sustained by recent developments. We are now seeking to include a range of different perspectives and new ways in which ethnographic displays are being produced, reshaped and reimagined.

Ethnographic museum displays have suffered from their historical association with imperial plunder and the subjection of colonised peoples, which means that in postcolonial society they are areas of debate, contestation and controversy. The early displays of the major ethnographic museums and galleries in the Anglo-American world were clearly cast within an imperialist mentality. Whilst German museums may have been less racist in their thinking and display (Penny 2002) there is no doubt that museums such as the Belgian Royal Museum of Central Africa represents vivid evidence of King Leopold's personal involvement in the history of the rapacious extraction of mineral and artistic wealth from the Congo (Hassett 2022). But, as Boris Wastiau explains in this volume, this was all part of the wider international network of imperial plunder. Recent critiques of such 'brutish museums' (Hicks 2020), with their display of colonial trophies in the heart of imperial cities, further reinforces this overwhelmingly negative picture. Demands to decolonise, repatriate and return objects seized during periods of asymmetric power have never been stronger – and ethnographic museums find themselves at the centre of these increasingly vocal calls.

But there are other reasons why ethnographic museum displays suffer such a negative reputation. Many of these relate to widespread intellectual confusion about their role and purpose. Both 'ethnographic' and 'museum display' are terms that are slippery and confused. Ethnography always requires a qualifier such as 'ethnographic realism', 'ethnography and archaeology', 'ethnography and evolutionary theory', and more profoundly, 'participant observation and comparison'. The word 'ethnographic' is valuable particularly as in its origins it asserts the importance of comparison. Similarly, museum display presumes either a typological approach, as is present in specialist collections such as textiles or ethnomusicology, or a narrative which incorporates a contextual element. When both these terms are juxtaposed the room for confusion and debate grows exponentially. There is a further question which becomes more insistent over time: whose ethnography are we considering – that of the colonisers, that of their subjects or a composite which is expressed in museum displays?

This book seeks to explore how individual case studies help clarify what is at stake. As the title implies, there are deep underlying historical problems associated with ethnographic display and these problems relate to political decisions about what is to be displayed and, just as important, what is to be excluded. We argue that only when these issues are faced can we return to the possibility of ethnographic display with a new confidence. We, like Thomas (2016), are convinced that new approaches can bring ethnographic museology back into a useful future.

The book is in three parts. The first (Chapters 1–3, 'The historic legacy of ethnographic museums') examines past ethnographic displays with the aim of discovering what can be learned from their history; the second (Chapters 4–8, 'Current practices') looks at contemporary display in terms of shifting practices and narratives; and the third (Chapters 9–12, 'Future directions') considers alternative ethnographic visions. Of course, none of these studies can be neatly compartmentalised – each study has its unique political and artistic history and common themes run throughout the volume. Nevertheless, we feel that broad trends and shifts of perspective can be detected over the past half century and more, and that new types of engagement can be discerned in both the achievements as well as in the recurrent failures of the curators and designers.

We approach the task in Part I by beginning with the first attempts by ethnographers to disassociate museum displays from the imperial past. A common theme in such approaches was the employment of the concept of multiculturalism, often expressed in the names of museums of 'world culture' in such locations as Edinburgh, Cologne, Liverpool, the Netherlands and Sweden, as well as several more worldwide. The first two chapters offer examples of how the process operated in practice. In Chapter 2, Burt explores the ways in which curators at the Museum of Mankind sought to engage with source communities in the creation of a series of national and regional displays. Usually this involved collecting material from the community both in Britain and abroad and involved community members in display decisions. Curators also employed dioramas to give an experience of the setting of, say, a Middle Eastern bazaar or a Yoruba compound. Despite widespread public endorsement, this approach died a sudden death when the museum was disbanded. Burt discusses the underlying reasons for this collapse. The second example is that of Birmingham's 'Gallery 33: A Meeting Ground of Cultures', discussed in Chapter 3 by Stanley. This permanent gallery achieved international recognition for its attempt to rethink the museum's historic collection and reframe it within contemporary concerns. As Stanley explains, a planning group with extensive community and academic membership played a very significant part in the shaping of the displays. Another innovation was the employment throughout the displays of an interrogative approach. So, the label on the gallery's Collectors' House asked 'Why did they collect? Do these artefacts present an accurate picture of the people who made and used them? Who decides what happens to artefacts when they go into museums?'. Gallery 33 had a very prominent design input from a commercial design agency that made the whole display look radically different to that in the rest of the museum. Initially well received, the gallery slowly faded and eventually expired with merely a whimper. It could be termed an example of display entropy. Stanley explores why such a fate could have occurred. Why both Gallery 33 and the Museum of Mankind remain important is that they represent examples of success but also ultimately of failure of an ideal. Failure is seldom explored in museum ethnography. Why? Perhaps because we have not created a vocabulary or set of criteria to evaluate museums' impact against their ambitions or because we have a very limited concept of what represents success against which to measure. Hopefully these two examples help explore failure afresh.

Part II, 'Current practices', offers a range of more recent and extant attempts to relinquish the imperial past and develop new approaches. The section begins with a troublesome imperial era museum and its seeking a new post-colonial display. In Chapter 4, Lismond-Mertes interviews Boris Wastiau, who shows how the Musée royal de l'Afrique centrale (now The AfricaMuseum), despite the involvement of Congolese citizens in its redesign, has failed to address the basic question of the past behaviour of Belgian entrepreneurs and military personnel in the super-exploitation of the people and their physical resources. Wastiau offers the example of a military trophy in the collection to show the problem facing the curators. But the redesign of the museum also has left some glaringly obvious racist visual reminders of the colonial past. The AfricaMuseum should perhaps be considered as at most a partially successful attempt to erase the past in its new displays.

The next two chapters explore how settler societies have wrestled within their museums to move on from their imperial genesis when their values clearly derived from the European past. In Chapter 5, Knowles charts the trajectory of Queensland Museum in Australia from the period of colonial settlement to becoming a junior partner in colonial rule in New Guinea and the Torres Straits Islands. She follows the recent actions taken to involve indigenous collaboration in the redisplay of all the Queensland Museum's collections. In Chapter 6, Mallon shows how, as an indigenous curator, he has become able in Te Papa Tongarewa to involve other Pacific Islanders and Aboriginal people in basic questions about what to collect and to display. This has involved a range of challenges and opportunities in moving on from a settler mentality to a more inclusive philosophy and practice.

The following two chapters introduce two major non-Western players in the development of ethnographic museum displays: Japan and China. In Chapter 7, Taku Iida notes how Japan has had an ambiguous role in ethnographic display, reflecting the political changes in its international status. In the late nineteenth century Japan was subject to considerable pressures from the great powers. But victories in the Sino-Japanese war in 1895 and the Russo-Japanese war of 1904–5, gave Japan a place at the imperial table at the Treaty of Portsmouth (1905), confirming its colonial power over parts of China, Korea and Taiwan. It was in this new context that Japanese ethnology developed, with an emphasis on folklore studies which extended to studying the cultures of the newly subject peoples. Taku Iida explores how the material culture of Japanese and colonial peoples was exhibited in Hoya Museum in Tokyo up until the end of the Second World War and how under American influence both the ethnography and display philosophies altered radically in the setting up of the National Museum of Ethnology in the 1970s.

In Chapter 8, Luo Pan discusses China's complicated relations with ethnology. At the founding of the Peoples' Republic the equal status and autonomous rights of all minority nationalities were recognised. These rights were enshrined in the 1949 constitution. The number and names of these 56 recognised nationalities became fixed by the State Nationalities Commission and their cultural practices were logged. So a form of state sponsored ethnography was established. It is within this framework that ethnographic representations in cultural parks and in displays in museums have taken place. Luo Pan discusses how, more recently, moves have taken place to develop a more nuanced and responsive approach to museum display. There remains the curious question raised by Sharon Macdonald (2017) as to why the National Museum of Ethnology, first planned in 1980 and now built, has not yet opened. Luo Pang addresses this issue directly.

In Chapter 9, the first of Part III of the book, Triastuti discusses how local ethnic identity in Indonesia has been reinforced by movements of young people, from the physical phenomenon of the national ethnographic museum and theme park (Taman Mini Indonesia Indah) to the blogosphere. Like China, Indonesia is very concerned to embrace the diversity of its peoples into a common nationalism. Taman Mini was the brainchild of the wife of the president of Indonesia, Ibu Tien Suharto, who spent five years creating the site. Taman Mini contains 27 *rumah adat* (customary houses or pavilions) and a range of other buildings to underline the state policy of *Pancasila* (unity in diversity). These buildings include an army museum, and churches and temples to underscore official ecumenism. Since its opening in 1977 officials, military and schoolchildren have been encouraged to visit the site to explore ethnic diversity but, crucially, also to recognise and celebrate their own ethnic and geographic identity. But with the decline in nationalism in the last decade, young people have been transferring their commitment to blogger communities which nevertheless remain based on ethnic identity. Triastuti explains how the Indonesian blogger communities can be seen as cultural artefacts in a new ethnography.

Recent developments in the discussion of repatriation of artefacts from imperial museums to the place they came from has intensified, particularly in the wake of Black Lives Matter and the Hollywood film *Black Panther*, that exposed structural racism and has led many museums to embark on processes of decolonisation. No more is this so than in West Africa, where plans are afoot to house such restored objects. Perhaps the most well-known is the forthcoming Edo Museum in Benin, Nigeria, where bronzes that had been looted from the city are now being returned by many international museums. But there are two other candidates seeking items from European museums – the Bët-bi Museum in Kaolack, Senegal, planned to open soon (at time of writing) and the Museum of Black Civilisation in Dakar, Senegal, sponsored and constructed by the Chinese government and opened in 2018. Perhaps the boldest claim to innovation is that made by the John Randle Centre for Yoruba Culture and History in Lagos. Its curator Will Rea explains its philosophy:

> One of the things that we wanted to do here was to interrogate museology as a construct and ask the question about why the western model doesn't work within the African context, and how we can create a space that isn't a museum in the traditional sense but is more like a theatre of living memory. (Rea, quoted in Kamali Dehghan 2024, 29)

In Chapter 10, De Jong questions whether restorative approaches are adequate solutions. He argues, instead, that museums in places like Jola village in southern Senegal have the potential to critique the post-colonial present by questioning the way in which decolonisation was conducted in the 1960s. De Jong explores the question of whether museums can provide objects with a stage to engage audiences in a future.

Finally, we consider alternative ethnographic visions. Following Harris and O'Hanlon's prediction of future developments in ethnographic museums, two chapters explore how digital technology has impacted on and offered new ways in which ethnography can be employed. Chapter 11 draws on the cross-disciplinary and cross-cultural perspectives of its authors – Clark, Coyne, Paterson and Shellam – and presents two recent collection case studies which focus on Western Australian biocultural material held at National Museums Scotland. It considers how collaboration between anthropologists, zoologists, botanists and indigenous knowledge custodians has the potential to mobilise biocultural collections for use in culturally appropriate and ecologically meaningful ways for the benefit of a variety of stakeholders. It also reveals these collections to be sites of Indigenous and non-Indigenous knowledge systems that can work together to generate holistic understandings of biocultural collections.

In Chapter 12, Were and Rowlands provide two examples of how digital technologies offer new forms of engagement and cultural restitution in 'new style' ethnographic museums. The authors' basic premise is that digital technologies may allow for a new economy of objects which sustains knowledge and revival practices without the need for the return of physical objects. Digital technology can ensure that the digital image can become an independent source of authority not inferior to a physical original. Were and Rowland foreground two case studies. The first is the Mobile Museum conducted at the Queensland Museum, as an example of digital facilitation of participatory design and digital return. The objective was to provide a web-based and mobile phone application for New Ireland people to access the Queensland Museum collections in order for them to revive cultural traditions. The second is the DigiDogon cultural resource. The central argument in this project is that museum collections created in the past by colonial collectors for mainly European museums now form 'object diasporas' that are 'potentially invaluable as a resource' for Dogon cultural activities in Mali. The digital collection now acts as an outreach tool to raise awareness of the heritage of Mali, and given its format, is very attractive to the young. Both examples open but by no means exhaust new possibilities for ethnographic museum displays and collections to breathe a new life, particularly in parts of the world where physical resources are limited.

Today there are exciting fresh developments in ethnographic display, and much of this innovation is taking place in West Africa. The Bët-bi Museum, the Museum of Black Civilisations, the Museum of West African Art and the John Randle Centre all intend to overthrow the traditional concept of the ethnographic museum through a new synthesis. They seek to recover artefacts of historic importance from Europe and the Americas to consolidate an authoritative African art history. They invite contributions from the Black diaspora, especially from the Americas, to extend their cultural references from a traditional past to include a cosmopolitan present. They also promote artists to produce work that envisages a fresh future, a 'theatre of living memory' (Rea, quoted in Kamali Dehghan 2024, 29), and a new form of discourse that reinvigorates the ethnographic museum project.

References

Aldrich, R. 2009. 'Colonial museums in a postcolonial Europe'. *African and Black Diaspora: An International Journal* 2 (2): 137–56. https://doi.org/10.1080/17528630902981118.

Altick, R. D. 1978. *The Shows of London*. Cambridge, MA: Belknap Press.

Bennett, T. 1995. *The Birth of the Museum: History, theory, politics*. London: Routledge.

Harris, C. and M. O'Hanlon. 2013. 'The future of the ethnographic museum'. *Anthropology Today* 29 (1): 8–12.

Hassett, D. 2022. 'Must Leopold fall? The renovation of the AfricaMuseum and Belgium's place in international debates on the decolonisation of public heritage'. In *Unfinished Histories: Empire and post-colonial resonance in Central Africa and Belgium*, edited by P. Fraiture, 43–62. Leuven: Leuven University Press.

Hicks, D. 2020. *The Brutish Museums: The Benin Bronzes, colonial violence and cultural restitution*. London: Pluto Press.

Kamali Dehghan, Saeed. 2024. 'Noise and colour: New Lagos museum aims to be a "theatre of living memory"'. *The Guardian*, 30 April. Accessed 21 August 2024. https://www.theguardian.com/global-development/2024/apr/30/john-randle-centre-yoruba-culture-lagos-nigeria.

Kirschenblatt-Gimblet, B. 1991. 'Objects of ethnography'. In *Exhibiting Cultures: The poetics and politics of museum display*, edited by I. Karp and S. D. Lavine, 386–443. Washington, DC: Smithsonian Institution Press.

Kuper, A. 2023. *The Museum of Other People: From colonial acquisitions to cosmopolitan exhibitions*. London: Profile Books.

Lidchi, H. and S. Allan (eds). 2020. *Dividing the Spoils: Perspectives on military collecting and the British Empire*. Manchester: Manchester University Press.

Macdonald, S. 2017. 'Why has China's National Museum of Ethnology still not been completed, decades since its founding? Or, why such a museum should be!' *CAMRAH Reflection*. Berlin: Centre for Anthropological Research on Museums and Heritage.

Penny, H. Glenn. 2002. *Objects of Culture: Ethnology and ethnographic museums in imperial Germany*. Chapel Hill: University of North Carolina Press.

Shelton, A. 1997. 'The future of museum ethnography'. *Journal of Museum Ethnography* 9: 33–48.

Thomas, N. 2016. 'We need ethnographic museums today – whatever you think of their history'. *Apollo Magazine*, 29 March, 1–10.

2

Ethnography and art at the British Museum's Museum of Mankind

Ben Burt

The British Museum is a legacy of an empire that collected 'curiosities' as an Enlightenment and colonial project to better understand the world. As its collections grew during the nineteenth century, they were split into Natural History and Antiquities, which was further divided by regional civilisations, until 'ethnography' emerged as a category for everything that was left over (see Burt 2019, 14–19 for a brief account of this history). Strictly speaking, ethnography, as the description of culture, should include all museum artefact collections, but the nineteenth-century category has persisted and so has the institutional distinction, despite attempts to normalise it as 'world cultures'. In 1910 the British Museum *Handbook to the Ethnographical Collections* described them as the products of 'savage and barbarous peoples', antecedents of more advanced civilisations. When Ethnography became a separate Department in 1946, its contents reflected this nineteenth-century world view by separating the peasants of North Africa and Asia from the Mediterranean and Oriental civilisations they belonged to and lumping them in with the indigenous peoples of Southeast Asia, sub-Saharan Africa, the Americas and Oceania (see Figure 2.1).

The British Museum had acquired its ethnography from both settler colonies, where indigenous peoples were removed or marginalised, such as most of the Americas, Australia and New Zealand, and commercial colonies such as most of Africa, Asia and Melanesia, where they contributed to the export of local commodities. When the colonised attempted to reclaim their cultural heritage from the later twentieth century, whether as indigenous minorities or governing majorities in their own countries, they had little influence on metropolitan museums and the British Museum was slow to respond to the global movement for

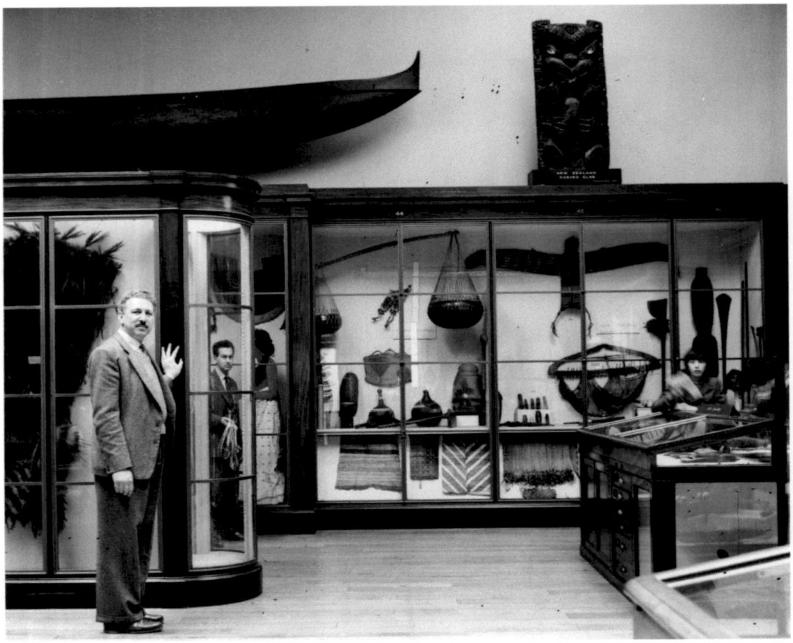

Figure 2.1 Part of the Oceania section of the British Museum ethnography galleries in 1968. Source: © The Trustees of the British Museum. Shared under a Creative Commons Attribution-NonCommercial-ShareAlike 4.0 International (CC BY-NC-SA 4.0) licence.

indigenous participation and self-representation in museum curation. But things began to change when the Department of Ethnography moved to the Museum of Mankind in 1970. The British Museum needed more space for its continually growing collections and Ethnography was large enough to make a difference as well as being marginal to its focus on Classical and Oriental civilisations. It would return when the British Library had moved out to a new site and made space.

As restored after the Second World War, the ethnography galleries at the British Museum had displayed the Department's collections in the kind of open storage usual at the time, crowded into grand glass cases organised by geographic area and labels with minimal background information.

The new Museum of Mankind staged temporary exhibitions carefully designed to attract and inform their audience by providing cultural contexts for the artefacts or emphasising their artistic qualities. Yet the 1910 *Handbook* was still used for public enquiries as the standard reference to the collections with no acknowledgement of its pervasive colonial racism, and Museum of Mankind leaflets were subtitled 'The British Museum's collections of Ethnography and Primitive Art'. The Museum exhibited a curious combination of colonial and humanistic values as the museum world moved towards 'decolonisation'.

The Keeper of Ethnography in 1970 was Bill Fagg, a respected Africanist art historian. He had overseen the reinstated ethnography galleries after the Second World War and now presided over new exhibitions in a Victorian building designed for the University of London in Burlington Gardens, a backstreet near Piccadilly. Large, plain lecture rooms were fitted out with floor and ceiling fixings on a grid system that supported exhibition cases, walls and partitions, constructed as required from aluminium joists supporting standard sizes of glass panes, blockboard panels and overhead light fittings. This flexible system was designed to facilitate a programme of changing exhibitions which Fagg and his senior curators, the Assistant Keepers, used to focus on particular geographical areas and cultural themes represented in their collections.

From the start, these exhibitions revealed a distinction between the 'Ethnography and Primitive Art' of the Museum's subtitle, often unacknowledged but sometimes explicit, especially when 'art' was concerned. Museum curators may criticise as 'pernicious' the tired and false choices between 'art' and 'ethnography', 'aesthetics' and 'contexts' (Shelton 2000; Thomas 2016), but these choices were and are made by those taking art historical approaches to exhibitions, at the Museum of Mankind as elsewhere. 'Primitive art' and its euphemistic successors ('tribal art', 'primal arts') are founded in the Western aesthetics of collecting and commerce that attempt to distinguish art from artefacts rather than treating one as a property of the other. An ethnographic approach acknowledges the artistic properties of artefacts within their cultural context, but the Museum of Mankind sometimes fell into the trap of attempting to raise the status of non-Western cultures by assimilating their artefacts to the prestigious category of 'art', rather than by explaining the context that gave them value within their societies of origin.

The distinction between ethnography and art was more obvious in some exhibitions than in others. Early exhibitions such as *Hunters and Gatherers: The Material Culture of the Nomadic Hadza of Tanzania* (1970–4) and *The Gonds of Central India* (1973–5) drew minimal attention to the artistic properties of the artefacts. Bryan Cranstone and Dorota Starzecka's *The Aborigines of Australia* (1972–82), *The Solomon Islanders* (1974–85) and *Hawaii* (1975–97) featured many artefacts that had been collected for their artistic qualities but arranged them according to social and cultural themes, that is to say ethnographically. Fagg's *Divine Kingship in Africa* (1970–3) built a pillared courtyard and earth altars to display brass plaques and sculptures as they might once have been seen in the king of Benin's palace in Nigeria, complemented by glass case displays. John Picton's *Yoruba Religious Cults* (1974–81) went further, placing ceremonial artefacts in alcoves around a cloistered courtyard as if stored when not in use, provoking criticism that their artistry was hard to see. These followed the tradition of museum dioramas dating back to the late nineteenth century, with a 1960s precedent at the British Museum when Bryan Cranstone displayed a collection from his expedition to Telefomin in the New Guinea Highlands in simple tableaux with enlarged field photos.

By contrast, Fagg's *The Tribal Image* (1970–1) displayed figure sculpture from around the world on freestanding plinths as in a European fine art gallery. Equally aesthetic but less focused was the longstanding *Treasures from the Ethnographic Collections* (1975–97), a changing display of artefacts in glass cases, the artistry of which apparently made any ethnographic context superfluous. Other explicitly art-focused exhibitions followed over the years but more combined art with ethnography. *The Tribal Eye* (1975–8) showed artefacts featured in a BBC television series of that name which, although focused on art, provided a rich cultural context in films of source communities around the world and enlarged photos for the exhibition. Other, mostly small, exhibitions focused on particular artistic traditions such as *Eskimo Sculpture, Malay Shadow Puppets, Bulgarian Village Arts, Maya Pottery and Sculpture* and *Ashanti Goldweights*, displaying them as art while treating them ethnographically as culturally situated.

What the Museum of Mankind became famous and popular for was its reconstructions, in which stage sets as context for the artefacts became exhibits in themselves. *Divine Kingship* and *Yoruba Religious Cults* were followed by Shelagh Weir's *Nomad and City* (1976–8), for which the ground floor galleries were occupied by a fully furnished Bedouin tent, leading to an urban street market and domestic interiors. The buildings were constructed by professional film set builders and incorporated many artefacts such as architectural woodwork, sacks of market produce and soft furnishings that were acquired, and kept, as accessories rather than as registered museum specimens. Critics compared the whole effect to a tourist experience, which was not meant as a compliment. An equally realistic effect was achieved by Brian Durrans's *Vasna: Inside an Indian Village* (1982–4), with a walk through a Gujarati weaver's house (see Figure 2.2) and a cart drawn by two fibreglass bullocks. In photos,

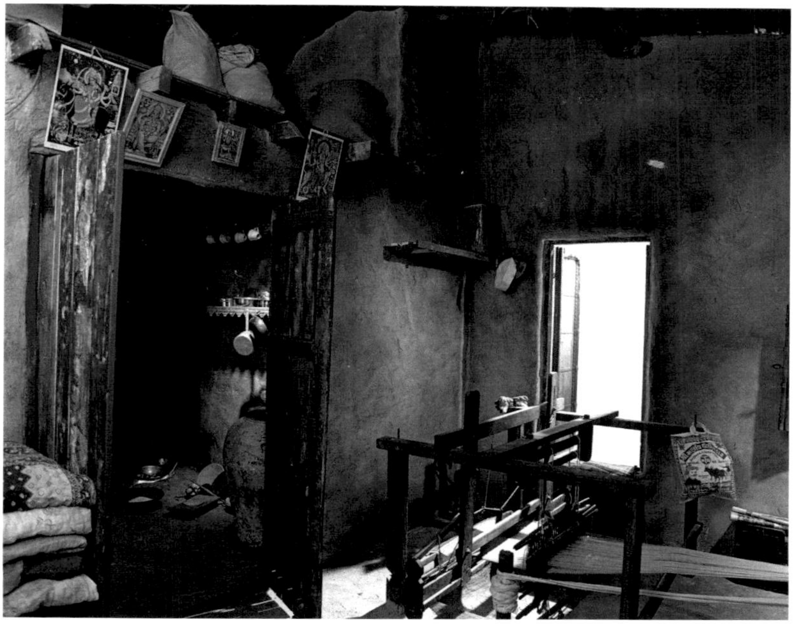

Figure 2.2 Inside the weaver's house in the *Vasna* exhibition at the Museum of Mankind, 1982–4. Source: © The Trustees of the British Museum. Shared under a Creative Commons Attribution-NonCommercial-ShareAlike 4.0 International (CC BY-NC-SA 4.0) licence.

these scenes were hard to distinguish from the real thing. Less elaborate but still impressive was John Mack's *Madagascar: Island of the Ancestors* (1986–8), which included scenes with two houses and a reconstructed tomb with memorial sculptures, imported with other exhibits from his collecting expedition.

All these exhibitions were motivated by a general desire to not merely promote appreciation of non-Western art and artefacts but also to provide understanding and respect for the cultures concerned. The curators appeared to acknowledge implicitly that 'ethnography' represented what popular culture and the British Museum still regarded as the inferior or second-rate cultures of the world and that these prejudices needed to be addressed. A few exhibitions had the explicit objective of challenging popular stereotypes and misconceptions. *Nomad and City* was part of a wider World of Islam Festival which provided sponsorship for the reconstructions and sought to promote a positive image of Arabia. Jonathan King's *Living Arctic* exhibition (1987–90), advised by anthropologist Hugh Brody, challenged colonial stereotypes by portraying the present day lives of Canadian First Nations through reconstructed dwellings in order to show their continuing close relationships with their land. The purpose of the exhibition, which led to sponsorship by indigenous organisations, was to advocate for the fur trade that maintained this way of life but was being challenged by anti-fur campaigns. Michael O'Hanlon's *Paradise* (1993–5), based on his field research in the Papua New Guinea Highlands, also sought to change conventional perspectives by showing how popular images of dancers in paint and feathers belonged to a society transformed since colonial contact in the 1930s into a contemporary peasantry engaged in coffee cultivation. Reconstructions of a store selling imported goods and a festival shrine were complemented by enlarged colour photos of spectacular ceremonial costume and painted landscape backdrops.

However, for many exhibitions cultural context did not extend to the history of the source communities. There was a strong tendency to present artefacts out of time in an 'ethnographic present' that left uninformed viewers to assume, by default, that they represented societies of the present day rather than their ancestors who had lost such things to collectors and museums. Even Dorota Starzecka and Jonathan King's *Captain Cook in the South Seas* (1979–80), which made a point of tracing the exhibits to their collection in Polynesia and the Northwest Coast of America, made no reference to the present day. *Asante: Kingdom of Gold* (1981–4), curated by Fagg's successor as Keeper, Malcolm McLeod, with reconstructions of a Ghanaian village, a palace façade and a tableau of a chief sitting in state, was created in response to Asante reparation claims for British conquest in 1900. Dealings with Asante were mediated by the Foreign Office, the exhibition was opened with great pageantry by the king of Asante, and it attracted many Ghanaian visitors. But it was set in an unacknowledged ethnographic present that did little to inform the British audience of this colonial history or its relation to Ghana eighty years on. In this it followed *Divine Kingship*, which made no mention of the British looting of the exhibits from Benin in 1897.

The ethnographic exhibition that failed most spectacularly to deal with colonial issues was *The Hidden Peoples of the Amazon* (1985–7), which exposed the underlying conservatism of the British Museum. A reconstruction of a present day Tukano communal house interior from Colombia with model people engaged in everyday activities was complemented by figures in ceremonial costume and artefacts in conventional glass cases, giving a vivid portrayal of Amazonian Indian life (see Figure 2.3). This was criticised by the campaign organisation Survival

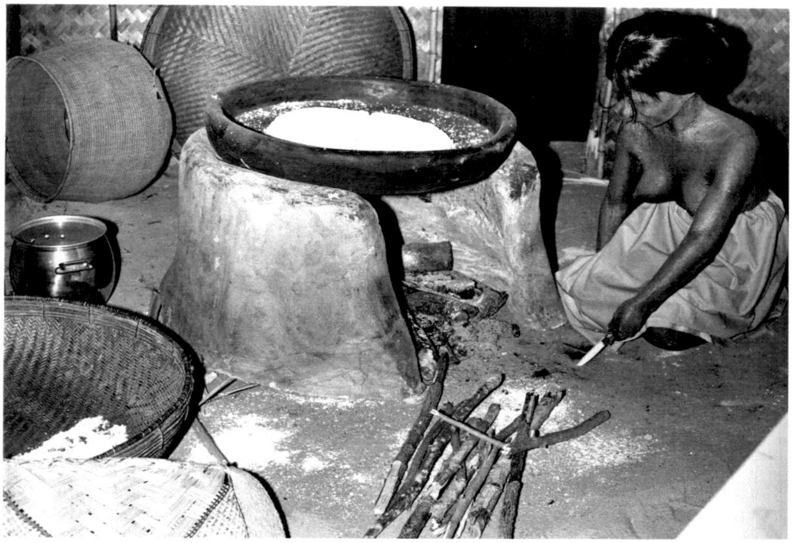

Figure 2.3 The cooking area in the *Hidden Peoples of the Amazon* house at the Museum of Mankind. Source: © The Trustees of the British Museum. Shared under a Creative Commons Attribution-NonCommercial-ShareAlike 4.0 International (CC BY-NC-SA 4.0) licence.

International for omitting to mention the colonial intrusions that were making this portrayal obsolete as well as ignoring the history of genocide and expropriation in other areas of the Amazon. Ironically, contributors to the exhibition included active supporters of Survival International, but the Museum refused the demand to represent this colonial context. The result was a public row, with Survival International picketing the exhibition. The whole episode reflected very badly on the Museum of Mankind and the British Museum.

Some art exhibitions did manage to engage with historical and contemporary issues. Brian Durrans's *Traffic Art* (1988–91) displayed a collection of painted panels from Bengali rickshaws in an art gallery setting, complemented by an exhibition with two rickshaws and several tableaux illustrating the hard work and lives of the rickshaw pullers. Shelagh Weir curated an art-historical exhibition of *Palestinian Costume* (1989–91), with clothing on mannequins but ethnographic context otherwise confined to backdrops of historic photos. This was a deliberate attempt to promote a positive identity for the source community by exhibiting an artistic tradition without emphasising the history of Israeli colonisation in a way that might provoke a Zionist political reaction that the British Museum would find threatening.

Eurocentric colonial attitudes were most obvious in some of the other art exhibitions. Any exhibition that presents non-Western artefacts as 'art' risks appropriating them to the Western culture of collecting and connoisseurship that has guided the plunder of artefacts from colonised lands throughout the colonial period and since. Artefacts gain prestige and commercial value as 'art' in a hierarchy of collectibles which is validated by experts in terms of authenticity (see Belk 1995). The Museum of Mankind curators seemed quite uncritical of the historical construction of the category of 'art' and its relation to bourgeois commercial collecting, although they were themselves implicated in it as professional collectors and academic experts. Bill Fagg was a leading expert in this respect and on retirement in 1973 became a primitive art consultant for art dealers Christies. Later curators continued this focus on African artefacts as art.

By the 1980s a new art history movement was reappraising the work of European artists inspired by the aesthetics of African and Oceanic sculpture in the early twentieth century. In 1985, while the Museum of Modern Art in New York staged *Primitivism in Twentieth Century Art*, the Museum of Mankind opened *Lost Magic Kingdoms* (1985–7), curated by the artist Eduardo Paolozzi (see Figure 2.4). *Primitivism* was criticised for exhibiting African and Oceanic sculpture to demonstrate its influence on European Modernism, rather than acknowledging it as art of value in its own right. *Lost Magic Kingdoms* repeated this theme by displaying artefacts from the Ethnography collections to complement Paolozzi's sculptures, as inspiration for his vision of exotic cultures. Rather than demonstrating formal affinities, as in *Primitivism*, Paolozzi used the artefacts to conjure up colonial stereotypes of 'savage and barbarous peoples' with no respect for their cultures and histories, juxtaposing his own work with artefacts such as trophy heads and religious sculptures.

Another art exhibition that lacked sensitivity to colonial issues was John Mack's *Images of Africa* (1990–3). This sought to challenge early prejudices about African culture by emphasising the artistic properties of Emil Torday's collection of Congo artefacts and the positive impression they made on the European art world at the beginning of the twentieth century. That was the time when loot from Benin was also leading academics to question the stereotypes of African savagery, but of course in neither case did it have much effect on European colonial attitudes of the time. *Images of Africa* also had the unfortunate effect of reaffirming the notion of 'darkest Africa' through illuminated cases of fine artefacts standing out against a background décor of dark jungle green. There was no mention of the particularly brutal Belgian colonial regime under which Torday made his collection.

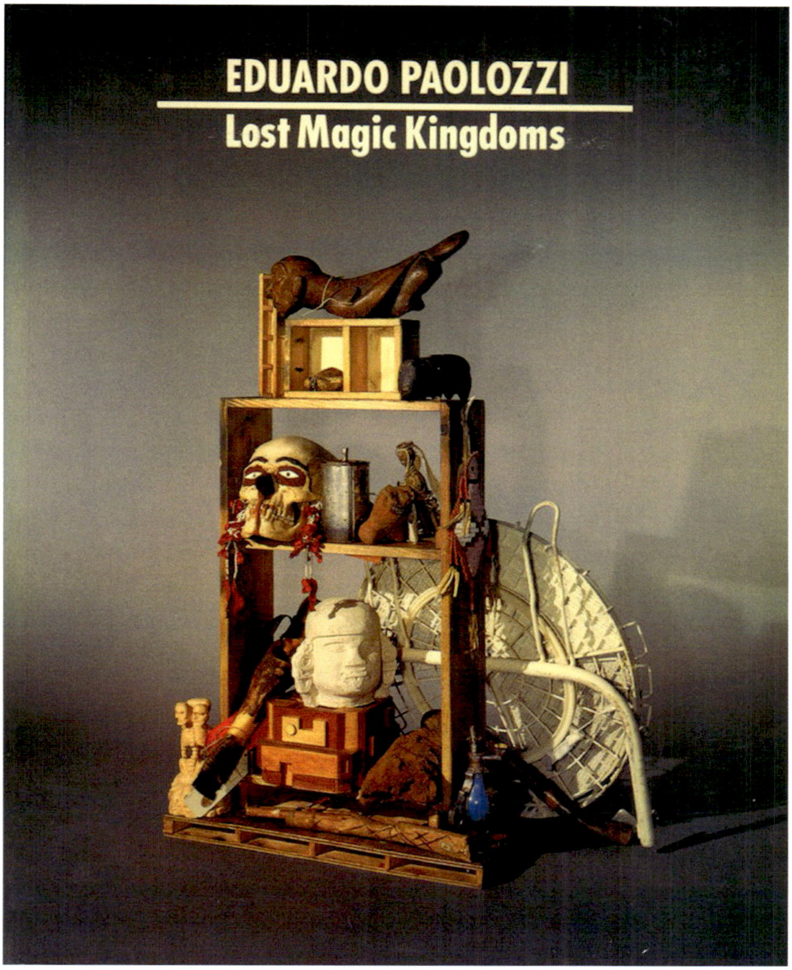

Figure 2.4 Eduardo Paolozzi *Lost Magic Kingdoms*, front cover of exhibition catalogue. Source: British Museum Press. Shared under a Creative Commons Attribution-NonCommercial-ShareAlike 4.0 International (CC BY-NC-SA 4.0) licence.

The British Museum's attitude to representing colonial appropriation and persecution was that this history was not really its concern as a museum and, as a national institution, it should not take positions on contentious political issues. It could not appear to promote campaigning organisations such as Survival International and curators who collaborated with them had to act with caution and discretion. It was out of the question for exhibitions to raise explicitly the kind of issues which have come to dominate debates about 'decolonising museums' in the twenty-first century.

However, this official position did not prevent the Museum of Mankind from exploring such issues through other means, particularly through the education service. Education, in terms of providing interpretation and instruction to teachers and students, was marginal to the British Museum's concerns, which were dominated by curatorial interests in research and exhibitions, and this also applied to the Museum of Mankind. As, at first, the only education officer in the Department from 1974, I saw it as my task to interpret the exhibitions for school visitors to complement, or compensate for, what had been presented by the curators. I produced worksheets for students and notes for teachers to provide the cultural and historical context that I felt to be missing from the exhibitions themselves, in language less academic than that used by the curators. I also assembled a Handling Collection of African artefacts that had never been incorporated into the registered collections and provided in-service training days for teachers and film and video shows for schools and general visitors. I was invited to work with various educational organisations that approached the Museum of Mankind for support with programmes and publications critical of colonial histories and development models, under what was often called at the time 'development education' and 'multiculturalism'. Although diverging from the Museum of Mankind's curatorial agenda, this was generally accepted as helping the public make better use of its resources, providing it did not make too many demands on the curators.

Then, in 1985, my colleague Penny Bateman was promoted to be my manager and, with the support the British Museum Head of Education, the two of us were able to expand the service at the Museum of Mankind, by organising teaching programmes. These began in 1987 with an *Arab Activity Programme* in a small unused gallery, furnished as an Arab living room where Arab teachers introduced their culture in conversation with a visiting school class. This was not tied to any current exhibition, but it set the pattern for a series of programmes supported by the funding raised for major exhibitions. *Living Arctic* employed a Canadian Inuk teacher for several months, who demonstrated his life at home with seal and caribou skins and clothing as props, which contributed to the Handling Collection. For *Palestinian Costume*, a dedicated Activity Room was built into the end of the exhibition, where a Palestinian teacher welcomed junior school students to her home and arranged for them to re-enact her old-fashioned wedding from years ago by researching in the exhibition and dressing in Palestinian costumes that had been donated to the Handling Collection. This was inspired by an educational trust, Learning Through Action, which was pioneering interactive programmes for schools and museums, using props and actors in dialogue with the students through role-play. The *Paradise* exhibition employed Learning Through Action to run a similar role-play programme in an Activity Room extended into the adjacent gallery, introducing development issues around coffee production in the Papua New Guinea Highlands.

By this time, Bateman's education initiatives were also having significant effects on the exhibition programme. She worked closely with Assistant Keeper Liz Carmichael to provide educational support for her *The Skeleton at the Feast* exhibition (1991–3) about the Mexican Day of the Dead, for which half of the upstairs galleries were occupied by altars for offerings to the dead, skeletal figures for decorating Mexican homes and fantastic papier-mâché skeletons including a tableau of a global apocalypse. This time the large mezzanine gallery became another Activity Room, open to the public to meet Mexicans demonstrating various crafts and for school groups to make and display such artefacts themselves. School programmes were run by a London educational group, Mexicolore, who introduced Mexican culture and history.

The education section even persuaded Nigel Barley, Assistant Keeper for West Africa, to curate a small exhibition on *Great Benin* (1993–7) to meet the demand by junior schools who had discovered Benin as a rare opportunity to teach about Africa through the National Curriculum. The exhibition ignored a consultation with teachers recommending historical context, but it gave us the opportunity to provide an interactive African Village programme. A West African educational trust, Heritage Ceramics, led by Tony Ogogo from Benin, used the former Activity Room of the Paradise exhibition. In a full day programme, they welcomed students to the village from the veranda of an African house, sent them to research the Benin exhibition, gave a workshop on pottery, textile dying or drumming, and concluded with everyone in African clothes taking roles in a village festival with drumming and dancing. Student responses included many letters with positive comments on life in Africa, contrasting as intended with the popular stereotypes of African poverty.

These programmes also contributed even more artefacts, especially costumes, to the Handling Collection, which came to be stored in the Activity Room and became a valuable resource for school and college visitors whose status could not justify access to the registered collections. It was a poor imitation of the teaching collection at the Horniman Museum which, as an educational institution, employed three teachers to use it for school programmes, but in some cases, it was all there was to offer as resources for the exhibitions. A case in point was *Images of Africa*, which made no allowance for an educational programme apart from a large cupboard in the gallery. There we kept a selection of African artefacts and a slide projector showing the colonial history and culture of the Congo, which were used by school classes with a pack of notes for the teachers and volunteer facilitators.

The school activity programmes tended to merge with programmes for the public, often aimed particularly at young families at weekends and school holidays. There were lively festivals at Eid for *Palestinian Costume* and the Day of the Dead for *The Skeleton at the Feast*, with performers and audiences from Palestine and Mexico, and African events with Heritage Ceramics. A British Museum interdepartmental exhibition, *Pottery in the Making* (1997) had demonstrations by various potters including Heritage Ceramics. Some exhibition curators raised funds for visiting craftworkers. Besides *Palestinian Costume* and *The Skeleton at the Feast*, these included one group of Native Americans carving and working textiles and another of painters and sculptors for a *Rain* exhibition (1996–7) on the arid southwest United States. Most remarkable of all were four Toraja men from Sulawesi in Indonesia who built a lavishly carved granary of wood and bamboo in one of the galleries (1987–96). Nigel Barley, straying from his remit as an Africa curator, recruited them and cared for them in his own home. Such visitors humanised the exhibitions, conversing with the visitors.

These education programmes had no problem with the ethnography–art dichotomy that affected some of the exhibitions. Their approach was to introduce peoples and cultures as contexts for the exhibitions and to focus students' attention on artefacts by drawing them or experimenting with practical artistic skills. But in the last years of the Museum of Mankind, until it closed to the public in 1997 in preparation for the return to the British Museum, the exhibitions themselves provided less opportunity for such programmes. From 1995, most of the galleries were given over to Africa '95, a festival at exhibition venues around London and beyond. Besides *Great Benin*, there were several exhibitions of African textiles and Chris Spring curated *The Power of the Hand* (1995–6), a large exhibition emphasising the artistic properties of African weaponry. There was also *Play and Display* (1995–6), reflecting the Africa curators' interest in African artefacts as 'contemporary art' in the European art gallery tradition. This featured welded steel sculptures of masquerade dancers from Kalabari in the Niger Delta, by British Nigerian artist Sokari Douglas Camp. In an irony unnoticed by the Africa curators, who had by this time eschewed contextual exhibitions as outmoded, this showed her works in their intended context of the art gallery.

The return to conventional art displays with less ethnographic focus was carried over to the new exhibitions that opened at the British Museum in anticipation of the return of the Ethnography Department. This was partly a matter of the cost of changing specialised exhibitions compared with permanent regional galleries behind glass, but it also reflected a change of curatorial attitudes, with contextual reconstructions falling out of fashion. The exhibitions began with *Mexico* in 1994, which followed the agenda of its sponsor, the Mexican government, in confining itself to the pre-conquest cultures of that country alone. It did not feature the meso-American region as a whole or include the heirs to those cultures, whether Indigenous or Mestizo, which were also well represented in our collections. Like a new Native North America gallery which opened in 1997, this reflected the colonial American perspective which appropriated Native culture and relegated it to the distant or early colonial past to affirm the hegemony of the conquerors and the marginalisation of the conquered (see Errington 1998, chapter 6). Both exhibitions displayed the artefacts artistically, *Mexico* with an authoritative archaeological commentary, North America with labels that paid no attention to colonial history. The North America gallery was titled, disingenuously, *First Peoples, First Contacts*, despite being dominated by artefacts made by people who had already been colonised.

In 2002, the next, much larger, exhibition, the *Sainsbury Africa Galleries*, continued the explicit art focus, as preferred by its curators Chris Spring, Nigel Barley and John Mack, and the sponsors, the wealthy supermarket family. Some of the gallery was arranged typologically, as textiles, pottery, metalwork, woodcarving and masks, there was a geographical focus for the Benin collection, and the introductory section was 'contemporary art' as produced for gallery display. Most of the gallery repeated the 1990s exhibitions at the Museum of Mankind, with minimal historical context nor any attempt to explain the vast differences between the regions of the continent or their links to Europe, Asia and the Americas. The objective of the Africa curators was to raise the profile and status of African culture to counter longstanding European racism, but their approach was counterproductive. Representing cultural and historical context through ethnographic exhibitions always took the risk of raising uncomfortable questions about colonial relationships, especially of misappropriation, as in the notorious case of Benin. The *Sainsbury Africa Galleries* evaded such controversy and instead validated African artefacts by assimilating them to the Western category of art, which legitimated the primitivist values of earlier generations still dominant in the art world of commercial collectors. Significantly, the galleries did not include the most prolific of new African artefacts, the commercial export sculptures that kept alive old traditions of woodcarving and brass casting by reproducing or reinterpreting masks and votive figures. These might at least have provided context for the contemporary art but collecting and exhibition policy applied criteria for authenticity drawn from the commercial values of the art world rather than the comparative discipline of anthropology. Export artefacts such as Benin brass castings and polished hardwood sculptures were dismissed as 'tourist' or 'airport art', unless they were dignified by age, like some Polynesian paddles and Native American argillite totem poles. An attempt to publicise the historical connections between Africa and other regions of the world as represented in the British Museum in an educational booklet (Burt 2005) was withdrawn by the Director for its mention of colonial looting.

At the British Museum, there were still temporary exhibitions that advocated the culture of source communities rather than art historical and academic appreciation. Dorota Starzecka's *Māori* (1998) responded to indigenous interest in the British Museum's exceptional New Zealand collections in collaboration with Māori scholars, artists and British residents. Colin McEwan's *Unknown Amazon* (2001–2) used atmospheric audio-visual effects to suggest a context for Indigenous culture and cosmology, drawing on the latest archaeological research. The Department's commitment to the cultures it represented was also evident in its continued engagement with source communities in public programmes organised with the education service in the Great Court, which opened in 2000. For *Māori*, there were indigenous artists working in the exhibition and public dance performances by the London Māori Club. For *Unknown Amazon* there were visiting craftworkers in a prominent position on the way to the exhibition. For Africa, there were several annual *Celebrating Africa* festivals sponsored by a Nigerian cultural broker, Bala Sanusi, for Black History Month. These filled the Great Court, with performances by visiting and resident musicians and dancers and stalls of arts and crafts, bringing African living culture, and visitors, to the heart of the British Museum. The schools programme, however, had to retreat from activity programmes as it came to depend on the British Museum's teachers who, in the absence of a dedicated Activity Room, held more conventional classes in seminar rooms and galleries. The Handling, now Teaching, Collection continued to expand and was used for such classes and for curatorial training for undergraduates, until it was rejected by a new Head of Education, fell into disuse, and was eventually moved offsite.

So, by the time the Ethnography Department staff and offices had relocated back to the British Museum in 2004 and the innovative education programmes were fading away, the art historical tendency had more-or-less seen off the contextual approach of ethnography. The Department was reorganised as Africa, Oceania and the Americas, with the Asian collections going to the Oriental Department, now Asia. Its main exhibition space became the former North Library, prominently sited next to the magnificent new Great Court. The gallery had to satisfy the ambiguous agenda of its sponsor, the Wellcome Trust, by dealing with medical issues. These were interpreted anthropologically as 'wellbeing' and became a series of ethnographic case studies from the various cultures represented in the Department, opening in 2003 as *Living and Dying*. Even so, the designer, who was also responsible for *Images of Africa* and the *Sainsbury Africa Galleries*, still ensured an artistic approach that made it hard for the curators to represent the cultural and historical contexts of their various sections.

It is now well over 20 years since the Museum of Mankind closed and the Ethnography Department was assimilated once again into the British Museum, with its exhibitions back behind glass, its education programmes in classrooms and its stores only now being moved back from an inconvenient site several miles away. At the time, there were hopes that the achievements of the Museum of Mankind might help to enrich and enliven the British Museum, but this failed to happen, for a number of reasons. The move was complicated by the expensive reorganisation of the spaces vacated by the British Library and the reconstruction of the Great Court, completed in 2000. Plans to build upon the Museum of Mankind's education and research services in a new store and study centre five minutes' walk from the British Museum were abandoned when that project failed due to funding problems. It was hard for Ethnography to prevail against the long-established domination of the antiquities departments in competition for the British Museum's scarce space and resources. Yet the failure of the Museum of Mankind to establish its innovative exhibition and education programmes at the British Museum also reflected the senior curators' abandonment of ethnography in favour of art historical approaches more in accord with the conservative culture of the British Museum.

Even so, the Museum of Mankind left an important legacy in contributing to the gradual shift in museum policy and practice towards the positive representation of its source communities and active engagement with them that now guides the British Museum and other ethnographic museums. Like the British Museum, the Museum of Mankind tended to fetishise its artefact collections, under the sway of the market-driven art world, to the neglect of its other cultural and historical resources, particularly its photographic archive. Yet it enabled its curators, as anthropologists, archaeologists and art historians, to research and publish according to their personal inclinations, which included sustained and positive relationships with the source communities and their advocates around the world. Despite inadequate exhibition space, the successor to the Department of Ethnography, Africa, Oceania and the Americas, has developed a strong agenda of active cultural research in collaboration with these communities. Significantly, the move back to the British Museum also reunited the marginalised 'savage and barbarous peoples' with the ancient civilisations that the British Museum had always been proud to display.

At the British Museum, as elsewhere, tensions remain between ethnography and art, between exhibiting exotic artefacts to communicate cultural and historical context and to express Western aesthetic and collecting values. Now that dioramic reconstructions have disappeared, curators have had to adapt their ethnographic perspectives to prevailing fashions for artistic exhibition design using graphic and audio-visual display. For all the curatorial criticism of the bogus art–ethnography distinction, there is no end in sight to the art-world colonisation of non-Western artefacts, or the need for 'ethnography' to contest it.[1]

Note

1 My thanks to Nick Stanley for his valuable editorial contributions. This chapter is based on my book, *The Museum of Mankind: Man and boy in the British Museum Ethnography Department* (Berghahn 2019), which elaborates on the history and issues mentioned here.

References

Belk, R. W. 1995. *Collecting in a Consumer Society*. London: Routledge.

British Museum. 1910. *Handbook to the Ethnographical Collections*. London: British Museum.

Burt, B. 2005. *Africa in the World: Past and present*. London: British Museum Press.

Burt, B. 2013. *World Art: An introduction to the art in artefacts*. London: Bloomsbury.

Burt, B. 2019. *The Museum of Mankind: Man and boy in the British Museum Ethnography Department*. Oxford: Berghahn.

Errington, S. 1998. *The Death of Authentic Primitive Art and Other Tales of Progress*. Berkeley: University of California Press.

Shelton, A. 2000. 'Curating African Worlds'. *Journal of Museum Ethnography* 12: 5–20.

Thomas, N. 2016. 'We need ethnographic museums today – whatever you think of their history'. *Apollo*, 29 March. Accessed 24 August 2024. https://www.apollo-magazine.com/we-need-ethnographic-museums-today-whatever-you-think-of-their-past/.

3

Success and failure: the life history of Birmingham's 'Gallery 33: A Meeting Ground of Cultures'

Nick Stanley

As a major manufacturing city at the beginning of the twentieth century, Birmingham had wide-ranging commercial relationships across the British Empire, and these were reflected in the city's museum collections. The ethnographic displays, first exhibited in 1930 in a new gallery, were taken down after the Second World War as British imperial ambitions waned. But the emergence of a new multicultural agenda in the 1960s enabled the museum to repurpose the abandoned ethnographic artefacts. The arrival of substantial numbers of new commonwealth citizens changed the demographic composition of the city, and the museum sought to respond with a gallery that would reflect this change, offering a 'meeting ground of cultures'. There were a number of strategies adopted. The principal aim was to create a revitalised ethnography, employing community-focused approaches that responded to the new populations. The development of the gallery was funded with major external support, and there was extensive preparation, including visits to international models and support through community and academic involvement, as well as the introduction of innovative design and interactive elements. There was a clear anthropological vision expressed in the organisation of the various displays, as well as a determination to reinvestigate historic ethnographic collections with fresh purpose. The gallery received extensive international interest and general approval for its daring innovations. As a result, the gallery survived for a quarter of a century. But, as it aged, its inherent contradictions appeared ever more starkly. This chapter examines the natural life cycle of this ethnographic gallery, in the hope of offering some more general criticisms that might have application in other modern and contemporary displays elsewhere. This chapter outlines how, despite new models of display, the thrust of multiculturalism is not a sufficient spur to overcome the problems of previous models. It also outlines how exhibition failures require serious consideration. This is part of a more general consideration of how failure is not only expected but also built into how knowledge and expertise are managed in a given social field (Fennell 2023, 192). Display theory must constantly grapple with this conundrum.

It seems that few curators, especially curators of ethnography, give much thought, as they go about creating new exhibitions, to Faust's dire warning, 'All that exists deserves to perish' (Engels 1886/1968, 597). At least, if they do, they certainly don't write about this prospect. This is a shame because if we neglect to consider the future, we lose the opportunity to gain experience from institutional and intellectual histories. When we discuss a new display, we seldom imagine the eventual fate that it will face as it is disassembled. So far, I have only located one notable exception, the article on 'lost museums' by Steven Lubar and his co-authors. They describe what they term museum taphonomy, the process by which collections disappear because of organisational decay and fossilisation. The authors give examples of why and how this might occur.

> Some [collections] are de-accessioned, deemed insufficiently mission-related, useful, authentic or not of the finest quality. Some are traded to other museums. Others, still of value, are claimed by disaster or the smaller tragedies of storage mishaps or misbehaving visitors ... Some are simply lost in the storeroom ... Finally, those objects still in museums can be lost in a variety of ways. In history and natural history museums, objects [can be] separated from the information about them. (Lubar et al. 2017, 2)

But, at the end of the day (and there is always an end to the day) museum displays die for three reasons: they become stagnant (often because they are not refreshed in any way from the original presentation), the display falls out of favour and dies from lack of love and/or the curator who generated the display leaves and the exhibit loses its champion.

If this all sounds abstract there are a number of exhibitionary candidates that can be examined in the context of taphonomy. One of the most frequently discussed is the Royal Ontario Museum's *Into the Heart of Africa* (Cannizzo 1989) where the intention of the curator was radically rejected by a part of its audience, the Canadian community of Afro-Caribbean inheritance. This disjuncture killed off the plan to tour the exhibition in North America and led later to an official apology from the museum to the local Afro-Caribbean community (Thompson 1989). At stake was the interpretation of historic imagery involving African people and Canadian colonisers. Henrietta Riegel notes that the crux of the problem lay in the employment of irony by the curator 'as a postmodern critique of colonial collecting practices' (Riegel 1996, 83).

Irony is the slipperiest of concepts and risks being missed or misunderstood, or unwelcome to viewers. In this case the result was the premature demise of the exhibition. At the same period a new ethnographic exhibition at the Field Museum in Chicago raised different, but equally problematic, questions. As John Terrell argues in his 'Disneyland and the future of museum anthropology' there was a persistent worry that 'we [Field Museum] were not keeping up with Disneyland and other contemporary forms of popular entertainment. There was the thought that something radical – new talent with fresh ideas from the outside – was needed' (Terrell 1991, 151). For the new permanent exhibition *Traveling the Pacific*, which opened in 1986, a well-known museum designer, Michael Spock was employed to radically change the Pacific displays. A major dispute erupted with curators over what should appear in the new show (an indicative noun), and interactive exhibits like a quaking volcano replaced the historic collections from Melanesia. A review of this dispute was entitled 'Say goodbye to the stuffed elephants' (Honan 1990). Spock was determined to overthrow curatorial hegemony. To begin his assault, he stated:

> Most people don't come here to be punished. If you're a family of five and you've been to the aquarium first, you don't want to face 200 headdresses; 20 is plenty. When confronted with a case filled with 200 carved ceremonial masks, except for convincing you that this was a culture with lots of free time to perfect these arts, you don't get any story, anything to follow ... throughout this exhibit, although there's lots of beautiful casework, there's no pacing, no chance to rest and too few modes of presentation. Finally, it's too dark, dismal and depressing. I call these galleries a magnificent failure. (Honan 1990)

The reaction of anthropologists was outrage expressed openly in journal reviews (for example, Kahn 1995; Rodman 1993). This dispute between curators and designers still rumbles on at the Field Museum, as in other galleries around the world, and what remains is a battle-to-the-death between what critical designers see as a vision offered by curators to visitors of a silent ahistoric portrayal of a people and their material world (and often with dioramas to situate them in the popular mind) and their opponents, the 'exhibit developers' who are keen not to overcomplicate a strong marketing image with anthropological verbiage. More recently, Le Musée du Quai Branly Jacques Chirac in Paris has raised similar disputes and conflict between two polarised views of 'museums of non-Western objects as scientific or aesthetic spaces' (Shelton 2009, 1). All three of these examples point to ways in which ethnographers and museum anthropologists have been outmanoeuvred by their critics. The question remains, however, as to whether these were tactical setbacks or a signal that deeper epistemological problems were involved. What we lack is a sense of the dynamics involved in the complex problematic involved in setting up ethnographic exhibitions. I think that a case study may help uncover some of these dynamics in both successful exhibitions as well as, inevitably, those that fail.

What I propose here is to offer an example, in some depth, to uncover as well as I can the variety of imponderables that surface in the making of an ethnographic exhibition, what motivates its creators, how it is conceived, put together, experienced and then how it may end. How did I come to select this example? Principally, I must admit, because I know it well, having been involved in its planning and construction, as will become clear in this account. 'Gallery 33: A Meeting Ground of Cultures' (henceforth Gallery 33) was a permanent exhibition at Birmingham Museum that opened in 1990 and lasted, intact, for 26 years. This study covers the evolution of the exhibition over time and its eventual demise.

Birmingham had extensive museum collections. From its creation in 1885, right up to the early 1930s, former West Midlands residents had sent their collections to the museum. Storage became a severe problem. The museum annual report complained: 'The need for adequate storage and exhibition space is becoming more acute every year, and it is practically impossible to display any recent presentations' (*Birmingham Post* 1928). Under pressure from the frequent arrival of these crates of artefacts, the museum decided to create an 'ethnological section' gallery, and to confine its remit to the South Pacific from where two collections were sourced, those from Arthur Wilkins and Ida Wench (*Birmingham Post* 1931). The display was offered as a piece of salvage ethnography; 'As their distinctive characteristics are fast disappearing in face of the onward march of Western civilisation, this collection must prove of increasing interest with the passing years' (*Birmingham Mail* 1931). The curator included in the exhibition 'illustrations of natives, customs etc., with descriptive matter, to give a sort of living interest to the exhibits'

(Peirson Jones 1992, 233). What we would now describe as racist tropes were prominent in the local press coverage. The mildest example was: 'From tomorrow onwards, visitors to Birmingham Museum and Art Gallery will have the opportunity to inspect articles which vividly depict the quaint customs and the ingenuity of the natives of the South Pacific area' (*Birmingham Gazette* 1931). The more sensational coverage was more in the garbled style of a colonial 'ripping yarn':

> The curio collection which he [Wilkins] made to the Birmingham Art Gallery and Museum [sic] includes skulls of head-hunters from the Sepik River in New Guinea, one of which originally surmounted the mighty and fighty [sic] frame of a big-boned Malekula chief [which] still has bloodstained hair on it. To procure such specimens he had more than once to take his life into his hands. (*Sunday Mercury* 1931)

After the end of the Second World War these types of display fell into disrepute. So, as Mary Bouquet notes, in the Netherlands, pre-war collections at the Institute of Cultural Anthropology, Utrecht University were taken down at the end of the 1960s. Such objects were either sent back to where they originated or retired to the museum stores while curators reflected on what to do with them next. This aspect of colonial history was thus forgotten or erased from collective memory (Bouquet 2001, 3).

The same happened in Birmingham. As Jane Peirson Jones, head of Ethnography (1975–96) noted, 'Ethnographic displays disappeared from museums in Britain in the 1950s and 1960s, out of embarrassment at the colonial focus which was out of place in a post-colonial world' (Peirson Jones and Stanley 2007, 71). But history was about to repeat itself in a new guise in the gallery that she opened in 1990. Wilkins and Wench were about to be recycled and emerge afresh. A critique of colonialism was about to develop using much of the historic collection of artefacts.

There were already straws in the wind. The Museum of Mankind, a department of the British Museum, started a new trend in 1970 – the re-invention of contextual displays 'where the artefacts formed part of reconstructed scenes of buildings and often people' (Burt 2019, 46). This was part of a movement to recontextualize collections through 'the study of what happens to objects, and to the people they attract, once they leave the hands of the original users, and particularly once they become appropriated by scholars, collectors, and museums in the wealthier nations' (Ames 1992, 46). The Museum of Mankind was also increasingly involved in community-based activities related to the displays, with the curator acting as facilitator rather than teacher.

Yet, the Museum of Mankind did not attract Peirson Jones and the team she built to construct the new gallery. Why? By 1990 the Museum of Mankind was something of a spent force, with its amateurism becoming increasingly obvious and embarrassing. As Clementine Delisse saw it, 'The D-I-Y approach to making exhibitions as an anthropologist was on its last legs. There was virtually no possibility any longer to position this amateur visual research against the authority of the interior designer' (Delisse 2020a, 135), and it 'felt both disconnected from movements in contemporary art or cultural studies, and out of synch with curatorial practice' (Delisse 2020b, 22). The Museum of Mankind had run out of steam at precisely the moment that it was about to be dissolved and reabsorbed into the main museum in Great Russell Street. One last heroic exhibition in 1993, *Paradise: Portraying the New Guinea Highlands* was successfully mounted despite having to employ the pre-existing exhibition set from an earlier show (O'Hanlon 1993, 84) but this was an experiment that could not to be repeated. The Museum closed at the same time as Gallery 33 opened.

There were other reasons that Gallery 33 eschewed the example of the Museum of Mankind. First, its creators had a different ambition and strategy. It was to be resolutely didactic and non-realistic. There was to be almost no place for dioramas or mocked-up sets. Instead, Peirson Jones looked to North American models, visiting a range of East Coast venues such as the Boston Children's Museum and especially the Smithsonian's Anacostia Community Museum (Peirson Jones 1991a). As a result, she became drawn into the Smithsonian's museum policy debates and contributed a chapter 'The colonial legacy and the community: the Gallery 33 project' to the Smithsonian's major review, published as *Museums and Communities* (Peirson Jones 1992). Contributors to this volume were engaged in the critical museology movement designed to 'sustain an on-going critical and dialectical dialogue that engenders constant self-reflective attitude towards museum practices and their wider constituencies' (Shelton 2013, 18). Furthermore, as the editors noted, 'Gallery 33 seeks to make current scholarship that is critical of the museum's part of the exhibit content' (Karp 1992, 155). This was to be achieved through clear and simple language whose brevity and strength would speak directly to the visitor. But the hallmark of Gallery 33 was its involvement of the audience through a resolutely interrogative approach. This involved 'a shift away from didacticism and the sense that the museum is a place where people can be stimulated and can acquire skills by asking questions such as "What is it?", "Why is it here?", "What is it for?"' (Thomas 2015, 43). Peirson Jones was convinced that questioning had to be at the heart of museum display, 'the only position we could take realistically was to raise questions about interpretation and to acknowledge that there are issues here – and to encourage the audience to explore these issues they've found' (Peirson Jones and Stanley 2007, 72). This questioning approach can be considered a form of what Macdonald calls 'estrangement' – 'a reflective technique primarily for throwing one's own presumptions into relief, and, as such, one that is deployed for raising questions rather than confirming expectations' (Macdonald 2020, 53). So, the first thing that met visitors on entering Gallery 33 was a panel on the front of the Collectors' House with a set of questions: 'Why did they collect? Do these artefacts present an accurate picture of the people who made them? What do people in those countries today feel about what happened to their heritage? Who decides what happens to artefacts when they go into museums? These are some of the questions raised by Gallery 33. There are no simple answers.' But there were some clear indications of where these questions would lead the engaged reader. Nothing was to be taken for granted, not even the displays themselves. All were to subject to rigorous evaluation by each individual visitor.

Unlike the Museum of Mankind, in Gallery 33 there was an investment in what Wayne Modest has called critical discomfort. As he explains:

> We've created an engine of the museum around happiness and comfort, as if it is a nice thing, so people should come out feeling good and happy. And it is not that people should come out feeling traumatised, but I see it as an investment in a critical discomfort. In particular, a kind of critical discomfort about the taken-for-granted-ness we have of ourselves (Modest 2020, 72)

Gallery 33 was designed as a challenge to everyday prejudice and careless thoughtlessness. It was also committed to what Ruth Phillips has termed a 'polyvocal model with multiple perspectives' (Phillips 2003, 163). This was a considerable intellectual step beyond the multiculturalism of the 1970s Museum of Mankind. Peirson Jones explained the genesis of the project in her Smithsonian contribution. She reported:

> The Director of the Museum, Michael Diamond set out four objectives for the exhibition: it should be unique in concept as

well as in quality of material; it should take account of multicultural issues specific to Birmingham; it should be designed to take account of the multidisciplinary potential of the museum service; and it should deal with public-interest issues rather than follow a traditional academic format. (Peirson Jones 1992, 222)

This sounds suspiciously like Peirson Jones's own agenda put into the director's mouth. Multiculturalism and multidisciplinarity expressed in everyday language were at the heart of the project. But the gallery's distinctiveness was to be made clear. Again, she wrote, 'I wanted to incorporate current thinking on the way cultural anthropology should be presented in museums, drawing on the experience gained in the South Pacific and in North America, where the growing politicisation of ethnic minority peoples has been most keenly expressed' (Peirson Jones 1992, 225). What this formulation did not address formally was the deeply unhappy situation that minority peoples in Birmingham, especially African Caribbean, were experiencing at the time. There had been riots in the Handsworth area of the city in July 1981, and in September 1985 when two shopkeepers were killed. These events were graphically recorded in John Akomfrah's powerful Channel 4 programme, *Handsworth Songs* in 1986. Violence was to erupt afresh in September 1991. In this charged atmosphere, the concept of inter-community tolerance was a constant feature in the planning of the exhibition.

The philosophy of the gallery appears initially quite simple:

> [W]e are all members of the human family, i.e., one race of people. We are separated by culture not biology … as an anthropological gallery in the 1990s Gallery 33 attempts to break with an exclusively Eurocentric perspective. It shows that all over the world people are involved in similar processes whatever the nature of their society. Gallery 33 can be a useful resource for multicultural education. (Peirson Jones 1987, 3)

The initial gallery brief began with a commitment, 'to represent the cultural heritage of some of the principal ethnic community groups in the city' (Peirson Jones 1987). But, perhaps surprisingly, Peirson Jones later stated equally firmly, 'It is not, as has sometimes been supposed, an exhibition about or indeed for Birmingham's ethnic minority communities' (Peirson Jones 1991a, 1). It was through careful examination of the collections that political lessons would be derived.

Peirson Jones went on to maintain that 'ethnographic artefacts represent a point in time where Black and white histories converge. The gallery therefore serves as a springboard for both multicultural exploration and race equality education'. She continued:

> [T]he ethnographical collection is itself a colonial construction and it has clearly been used in the 1850s and 1930s presentations to support a racist ideology. It therefore seemed appropriate to explore this by examining how the collection came to be formed and how they [sic] have been used in the museum process. (Peirson Jones 1991b)

At the heart of the enterprise there remained an ambiguity about the celebration of ethnic heritage and multicultural development. How was the conundrum to be resolved?

The answer was through juxtaposition, 'the familiar with the unfamiliar, the past with the present and the majority with the minority', as Peirson Jones wrote, continuing, 'I hope that the discordance created by these unfamiliar juxtapositions will attract visitors' attention and challenge a sense of order, their sense of the "other", and thus their sense of themselves' (Peirson Jones 1992, 227). Juxtaposition was to be a physical reality brought about through the design process. Peirson Jones was keenly aware of the Smithsonian emphasis on the crucial role of design in display. As Karp and Lavine argued, 'We need experiments in exhibition design that try to present multiple perspectives or admit the highly contingent nature of the interpretations offered' (Karp and Lavine 1993, 7). The Collectors' House had pride of place in the gallery as a prominent location for such experimentation (Figure 3.1).

Cut-out models of the three collectors, Arthur Wilkins, Percy Amoury Talbot and Ida Wench, foregrounded in the 1930s display, appeared on a mock-up of the veranda of Ida Wench's house in the Solomon Islands. Through the window, visitors could see an array of artefacts from each collection. What visitors would not have recognised is that these objects were mixed with others collected very recently. To gain an appreciation of the display required the visitor to engage with the interactive video that sat alongside the house. Without that gloss, it remained a curious and strange relic of historic figures in a gallery that otherwise looked like a 1980s illustrated cultural anthropology course. Despite Peirson Jones's intention that this group portrait should be nostalgic and emotive, raising questions about their motivations and selection criteria, there were precious few clues in the gallery to its

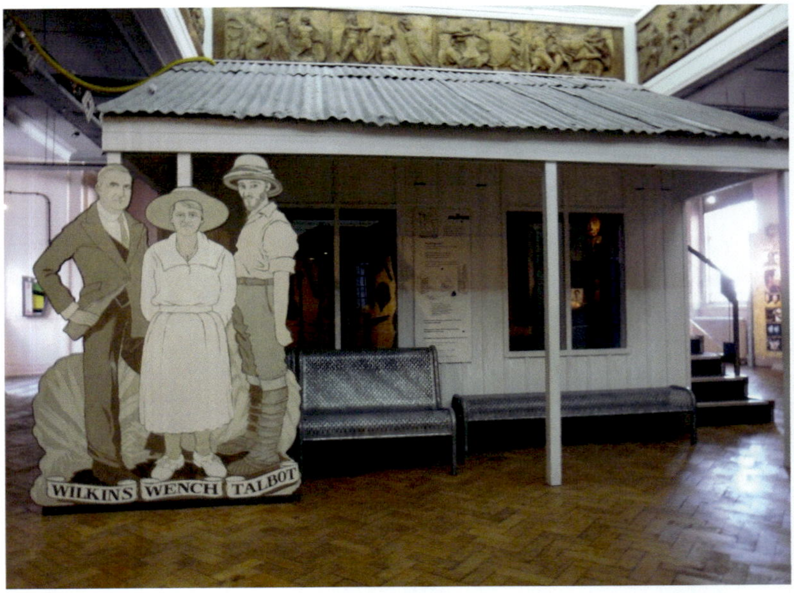

Figure 3.1 The Collectors' House display at Gallery 33.
Source: © Nick Stanley.

purpose. The period costume suggested the actors' historic time but the link with their artefacts had to be explained in a different way.

One way to provide historical linkage was through creating a contemporary fresh perspective. I had a part in this. I was sent to locate Ida Wench's mission station in Gela Province, Solomon Islands to interview her students, and make a contemporary collection. The students, by then in their eighties, decided that what I was seeking was a memento of Wench, so they took me into a clearing in the bush where they had uncovered a large cast-iron bath made in Bilston. This was indeed a potent relic, reminiscent of a colonial lifestyle, and clearly a powerful link to their engagement with her. My contemporary collection came, more prosaically, principally from the Solomon Islands National Museum shop.

Gallery 33 was to be distinctive in further ways. It was to have a hands-on component to be delivered via interactive programmes which sought to draw the viewer into the questioning approach to objects and concepts that was pervasive in the display script. A Phillips/MCA LaserVision video installation, entitled 'Collectors in the South Pacific'[1], had a crucial role in connecting the historic collections to a contemporary setting. As Peirson Jones recorded:

> The collective interactive video was developed to support this installation. The aim is to present a well-documented resource which exposes different viewpoints on related issues such as mission activity, cultural property and the role of museums in economic development. Visitors can explore these issues for themselves and join the dialogue. (Peirson Jones 1990, 2)

One of the principal attractions of the format was that it enabled multiple perspectives to be presented simultaneously. 'It is possible to convey counterpoint and conflict as well as personality and mood, quickly and effectively' (Peirson Jones 1990, 2). The interactive video had a lot of work to accomplish via its four interlocutors: Ida Wench, the missionary and Arthur Wilkins, the businessman/collector as historic figures; N. S. Jones, an American tourist (using text created by me) and Lawrence Fonana'ota, the curator of the Solomon Islands National Museum as contemporaries. In the installation, each of the narrators has a script relating to their own role, but they all comment on the changing role of objects over the past century. So, for example, bags that were historically made from pandanus and other fibres are now made often from nylon and other plastics which refuse to degrade and so pollute the Pacific. Fishing has likewise been transformed from individual endeavour in coastal canoes to the mega-trawlers out of Japan and the United States. What threat does this present for the future of fishing stocks for both Pacific Islanders and for the wider world? Historic photographs taken by Ida Wench are contrasted with the bustle of modern Honiara. Viewers of the video are asked a series of questions such as: 'Should objects like these in museums be returned to their countries of origin? Of what importance are modern versions of fishing gear in the preservation of cultural values from the past?' Visitors were invited to choose which person to meet and engage with or, alternatively, go directly into the database of 117 objects that was presented alongside.

The interactive video had the virtue of conciseness. As its creator, New Media Productions boasted, 'using a three-dimensional display to communicate the storyline of Collectors in the South Pacific would have taken up most of Gallery 33' (Peirson Jones 1990, 7). But, as will become apparent later, this was to be a double-edged sword. If the gallery embraced the most contemporary of media, it was similarly wedded to advanced interior design development. Peirson Jones stated emphatically: 'I wanted the distinctive conceptual approach to be reinforced by a design that would create a space very different from any other part of the museum' (Peirson Jones 1992, 225). The design brief stipulated 'a high-tech design and colour scheme and logo inspired by the mural art of Ndebele women from South Africa' (Peirson Jones 1992, 225). The design investment went further. As an early reviewer commented:

> Although there is no attempt to disguise the basic Edwardian structure, fittings are resolutely high-tech with exposed services (including controlled humidity piped directly into the showcases) and much use of lightweight metal structures. 'There's a metal theme going through it', says Michael Orr, one of the firm's partners. 'It gives a lightness of structure and interesting textures which don't compete with the collection'. (Grimley 1990)

Another reviewer was less enthusiastic:

> In contrast to the colourful entrance, the remainder of the gallery is grey and harsh in its decor. Many of the display panels and stands are made from steel mesh, metal piping and wood chip boarding. The result is cold and mechanical, with a modern hi-tech or industrial feel. (Simpson 1992, 74)

Others commented on the mismatch between the austerity of the fittings with the fussiness of the cabinet designs.

Peirson Jones recognised that her ambition was controversial but dismissed these criticisms airily: 'The high-tech visual presentation and the dominance of metals in the exhibition design have been questioned. These are partly a designer signature, and the response is a matter of personal taste. However, the use of metal supports and finishes can be seen as an ironic reflection of the source of Birmingham's wealth and patronage, which was based on worldwide trade in metals' (1992, 217). Perhaps irony was a risky idiom to employ. She was indeed successful in her plan to create a contrast with the rest of the museum, but the design decisions were to divide museum staff into supporters of the scheme and quite a sizeable group who hated the irruption of an alien presence into the museum's peace and quiet. This latter group may have taken consolation from the fact that Gallery 33 was at the very end of the circulation path of the museum and that 80 per cent of those that found the gallery did so serendipitously (Peirson Jones et al. 1993, 10).

The contrast with the rest of the museum was philosophical as well as practical and visitors were to be challenged by the displays.

> Museums are places where 'things' are usually classified and separated. Gallery 33 declassifies and mixes the familiar with the unfamiliar, the past with the present, and the majority with the minority. I hope that the discordance created by these unfamiliar juxtapositions will attract the visitors' attention and challenge their sense of order, their sense of the 'other', and thus their sense of themselves. (Peirson Jones 1992, 227–8)

As a reviewer commented:

> Unlike most other anthropology galleries which are arranged in typological order, or by geographical or cultural group, Gallery 33 is thematic in its approach. In this way it attempts to show the similarities which exist between the artefacts and practices of peoples in different parts of the world. (Simpson 1992, 72)

Here, Peirson Jones was endorsing a maxim proposed by Michael Baxandall:

> Exhibitions in which different cultures are combined or juxtaposed are inherently more wholesome than exhibitions of single cultures. The juxtaposition of objects from different cultural systems signals to the viewer not only the variety of such systems but the cultural relativity of his own concepts and values. (Baxandall 1991, 40)

Yet, beyond the Collectors' House the gallery proper looked less controversial in its employment of vitrines, or what Michael Ames ominously calls 'glass boxes' (Ames 1992, 140). See Figures 3.2–3.5.

Seven glass cases displayed artefacts relating to the following topics: music, signs and symbols, eating and drinking, body decoration, masks, African textiles, and societies.

The 'Societies' case was the largest and did the most anthropological instruction. It dealt with seven aspects: gender, society, politics, rank, ethnic identity, life cycle and religion (Figure 3.6).

A photomontage of the faces of citizens of Birmingham made by Vanley Burke covered one wall and, of course, the Collectors' House dominated the front of the gallery. Each display was interrogative in its orientation. So, the visitor is asked: 'what is society?', which elicits the answer, 'a society is a group of people who depend on each other and have their own customs and patterns of behaviour. These are based on common beliefs and values'. This panel leads into ethnic identity.

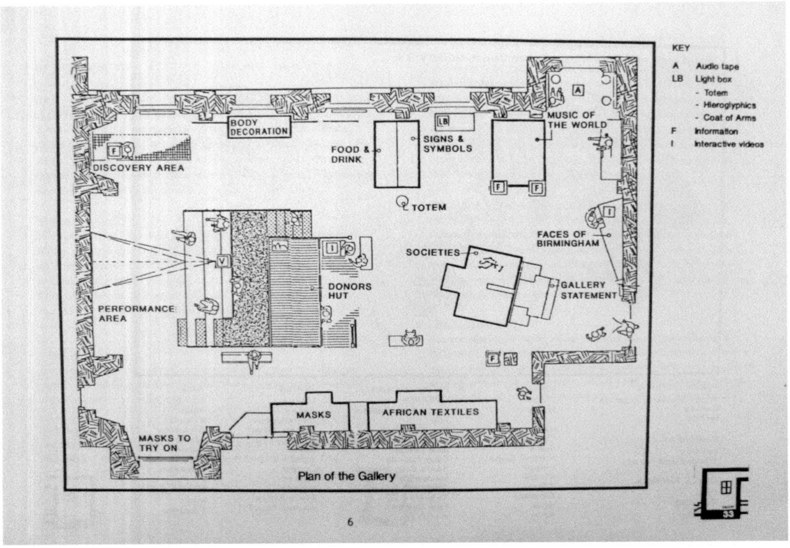

Figure 3.2 Plan of Gallery 33. Source: Birmingham Museum. © photograph Nick Stanley.

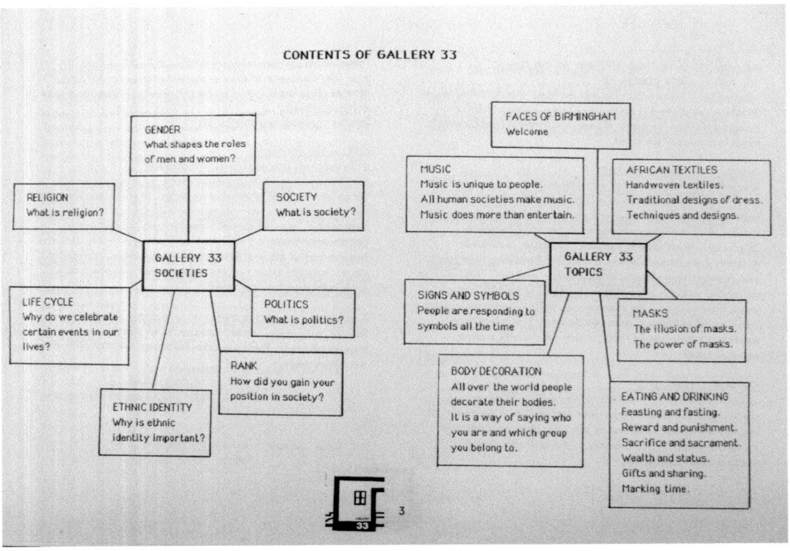

Figure 3.3 Plans for the 'Societies' and 'Topics' displays in Gallery 33, as spider diagrams. Source: Birmingham Museum. © photograph Nick Stanley.

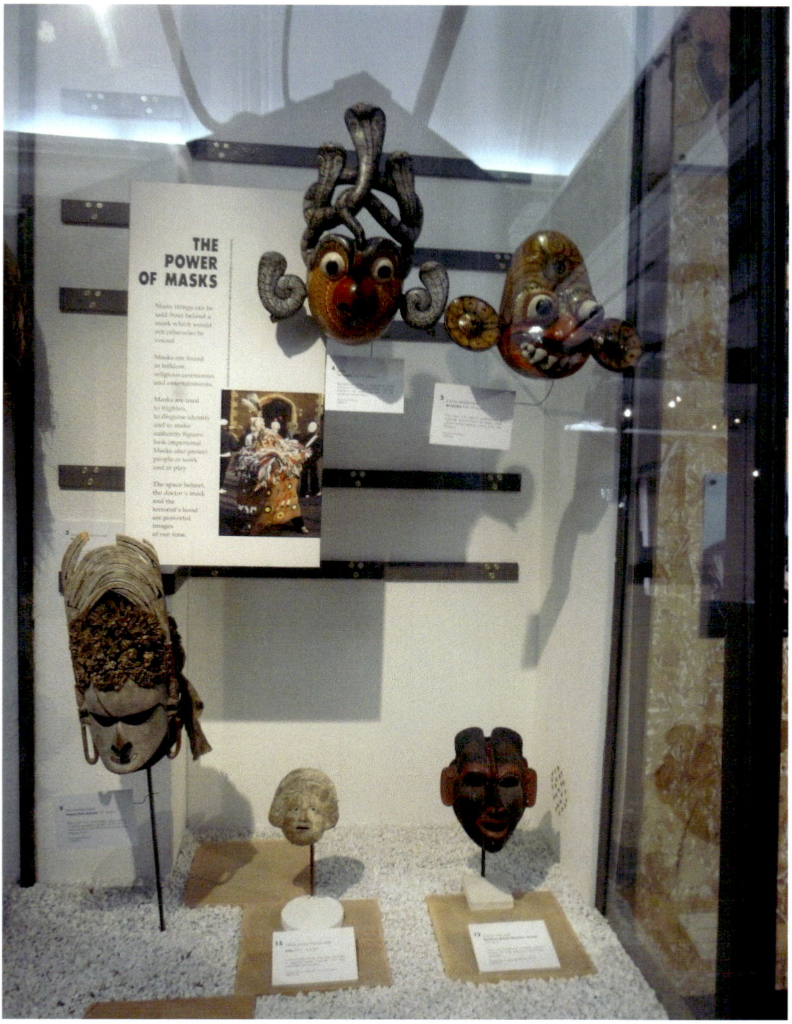

Figure 3.4 'The Power of Masks' display in Gallery 33.
Source: © Nick Stanley.

Again, the question arises: 'why is ethnic identity important?' The answer is that 'strong ethnic identity holds societies together. It adds up to a sense of self-worth and belonging which everyone needs'. The topic is illustrated with a photograph of 'Brummies shopping' and five human figures from different parts of the world. These include a Tabwa figure from Central Africa, a Mohave doll from North America, a bronze figurine from France c. 1900, a figure from the Huon Gulf, Papua New Guinea

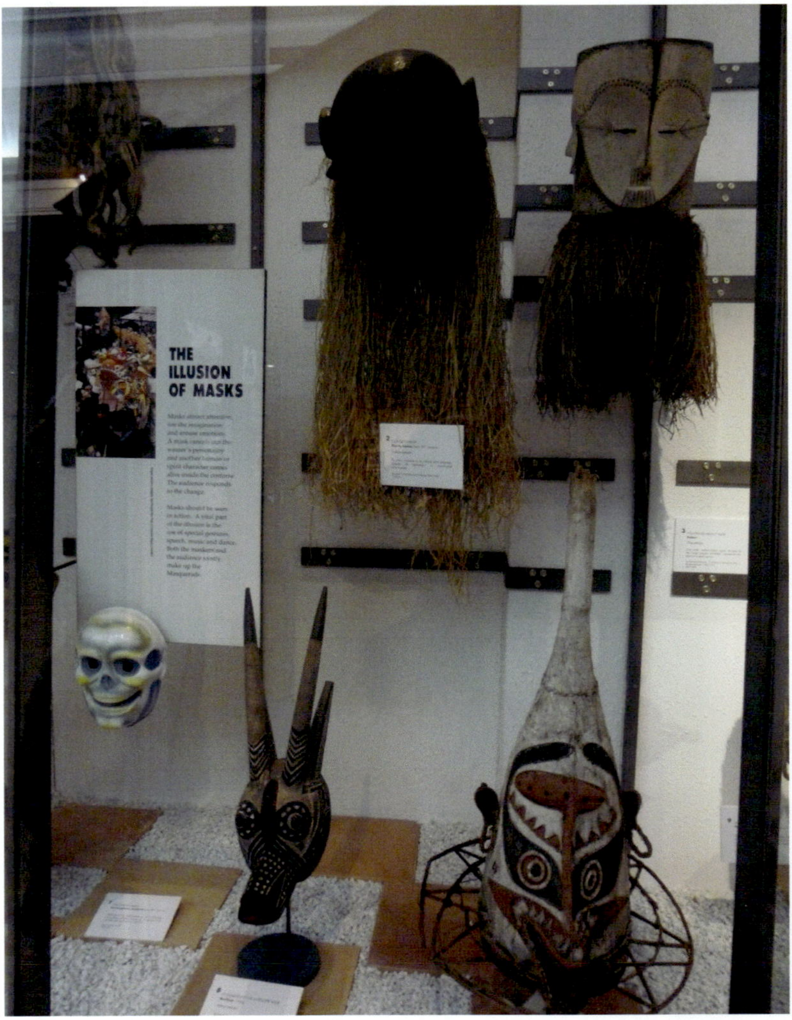

Figure 3.5 'The Illusion of Masks' display in Gallery 33.
Source: © Nick Stanley.

and a Japanese doll. A similar approach appears in each of the other cases. In 'Rank' the question posed is 'how did you gain your position in society?' The photograph is of Queen Elizabeth II at a state banquet for King Olaf of Norway and the six objects are a ceremonial club from New Caledonia, a 'slave document' from Litchfield, a breast ornament from Fiji, a signed photograph of Michael Jackson, a sacred threat from India and a British conspicuous bravery medal. The photograph in the Politics

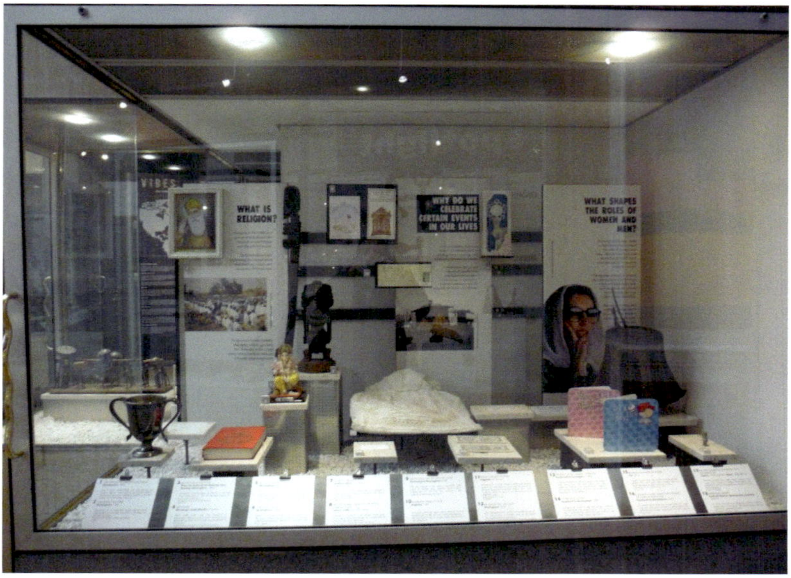

Figure 3.6 'Societies: Religion, Life cycle, Gender' display in Gallery 33.
Source: © Nick Stanley.

case is of President Gorbachev, President Reagan and President-elect Bush in New York in 1988 and among the objects are a chief's stool from Kenya and a copy of Hansard. The Gender case photograph is of Benazir Bhutto during the election campaign in Pakistan in 1988.

In all, over 500 objects were displayed, many of them quite large. Peirson Jones's argument was that

> ... the artifact ceases to be a relic of an exotic past, or a work of art, or simply a piece of loot. Instead, it becomes symbolic of complex colonial and postcolonial relationships, and can inform our understanding of the present-day world. (Peirson Jones 1992, 235)

What the 200 labels attached to the artefacts in the cases achieve is, to follow Holbraad's injunction, to submit the constellation of objects in each case to an overarching conceptual framework, in other words to treat concepts as things (Holbraad 2011, 12). Behind Peirson Jones's apparently anthropologically bland smorgasbord was an attempt to advance ethnic heritage interpretation in museums by exploring the concepts of power, subjectivity, artefacts, representation and facilitation (Peirson Jones 1991a, 7). The power of curators in creating the display has to be acknowledged, she argued, as well as the subjectivity that they bring to the work, the under-representation of peoples in displays, and the opportunities that need to be seized to remedy these disjunctures.

A major imponderable facing all museum displays is their effectiveness in meeting their objectives. It is notoriously difficult to elicit meaningful responses from the public, but audience surveys have often attempted to meet the challenge. Gallery 33 was no exception. Peirson Jones was keen to hear about visitor experiences and to evaluate their engagement. She introduced the visitor survey.

> Since the project involved new forms of experience for museum staff, a critical evaluation of the outcomes had become a necessary part of professional and institutional development. Gallery 33 attracted wide-ranging professional interest and is used as a text in a number of undergraduate and postgraduate courses. For these reasons it was appropriate to make the evaluation available to a wider audience through full publication. (Peirson Jones et al. 1993, 5)

Overall, the eight chapters recorded a high degree of satisfaction, but some significant issues were faced. The Collectors' House was largely ignored by visitors despite its prominence. The interactive video station was too small, and many people did not get a chance to engage with it when a small group of visitors had 'captured' it for a lengthy period. But the main problem, repeated over and over again, was the isolation of the gallery from the main body of the museum. The fact that such a high percentage of visitors discovered the gallery through accident meant that signage and directions were either negligent or, perhaps, evidence of an ambiguous relationship between the gallery and the administration. If the location of the gallery had been nearer the entrance its fate might have been very different.

Who did visit and comment on their experience? A broad range of ages, but a very white population (86%), had spent quite lengthy periods looking evenly across the different presentations. Perhaps, rather obviously, other gallery curators and students had the most to say. Chris Wingfield commented:

> One of the interesting things about Gallery 33 was that I regularly got letters, emails from students who were doing masters dissertations and essays. They had heard about it and wanted to know what happened. They wanted to use it as a case study. That partly pricked my interest in it. It also foreshadowed a whole series of developments

> in museums in the 90s. It did a lot of things first in a British context that were picked up by other museums which was partly why the students were interested in it. I remember reading comments from them. It seemed that it was a path-breaking gallery that had done lots of things right. (Wingfield, interviewed by author[2])

But who did not visit? Members of the Afro-Caribbean community, who, Peirson Jones observed ruefully, regarded the whole idea of ethnography and museums as decidedly Eurocentric and not worth embracing.

Jane Peirson Jones left the museum in 1996. She was replaced by Lisa Harris and then in 2004 by Chris Wingfield. Although he only stayed for three years, he took it upon himself to review both the successes and the shortcomings of the gallery over its 15-year history. He declared it a success in meeting its primary objective – its celebration of the common heritage of humankind exhibited in the variety of objects that have been employed to meet common human needs. The principal change that Wingfield made was to replace the African textiles display, which had been exposed to the light for far too long, with one from the George Grenfell collection made in the Congo and held by the Baptist Missionary Society which entered the museum in 2000. Wingfield also updated the gallery through the replacement of the Music display with a display called 'Vibes' which was about the roots of black music and covered West Africa, the West Indies and Britain. There was also, at one time, a proposal that the Collectors' House be replaced with a Caribbean beach bar, but Wingfield was concerned that this would further fragment the narrative of Gallery 33 (Wingfield, interviewed by author). Despite other minor changes, the task of reshaping the gallery was just too daunting to undertake without further funding. So, it stayed almost unchanged for a further decade.

Wingfield reflected on the extant gallery and identified three serious shortcomings which had reduced its effectiveness. The first of these was the promotion of novelty in the display. This was of two kinds. First was the determination to be up-to-the moment in looks, 'the many bright colours, off-centre angles, and a major use of industrial metal' (Wingfield 2006, 33) as well as contemporary photographs of world leaders. These served to date the exhibition badly and it was never easy to update these elements successfully. But the more fundamental problem in this innovatory display was the employment of new technology, in this case the videodisc video station. As Fernando Rubio has argued:

> Objects are fragile and temporal realities. Rarely, if ever, do these approaches take into account the fact that objects wear down and

> change, that they break, malfunction, and have to be constantly mended, retrofitted and repurposed, or that they are routinely misused, misrecognized and disobeyed. (Rubio 2016, 60)

The paradox here is that the most technically sophisticated objects like interactives (artefacts in themselves) are the most prone to break down and disappear. We spend inordinate amounts of time and energy trying to keep them working only to discover that they have become obsolete and unavailable. Appadurai and Alexander remind us that 'The proposition that technology is always effective, if only its users were not so fallible is a "solutionism" forever trying to solve technological limitations of malfunctions by investing more capital in designing new technologies' (Appadurai and Alexander 2020, 2).

This was the fate of the interactive videodisc installation 'Collectors in the South Pacific'. After three replacements, the makers were no longer able to repair the machines, which became a mute reminder of their obsolescence in the gallery. Rubio argues for 'the need to take seriously the temporality and fragility of the material world we study. Taking these things seriously means accepting that we cannot take *any* object for granted, even the seemingly unassailable ones' (Rubio 2016, 81). The failure of the videodisc destroyed the coherence of the gallery.

Wingfield's second problem with Gallery 33 was its chirpy indifference to its location within the museum. It appeared to be self-contained, which had both positive and negative consequences, and yet it retained its classical frieze of low-relief sculptures around its roof, creating a visual discordance that was hard to ignore. Perhaps the most important of Wingfield's strictures was his complaint that insufficient attention had been paid to succession planning. The gallery had required outside funding of £250,000 to get it running, and yet no budget had been raised to sustain it over time. After the initial costs, Peirson Jones had been aware, it would have been difficult to make changes once the gallery was running (Peirson Jones and Stanley 2007, 70) but no further funding was forthcoming from the museum: the gallery was left to fend for itself. Wingfield moved to a new post in 2006 and was replaced by Adam Jaffer who made few changes to it, concentrating instead on a new gallery of religious faith in the city (Bridgman and Jaffer 2017). He felt that Gallery 33 had run its natural course and proposed closing it in October 2016. I was working within the archives in the museum basement on 16 May that year when I experienced what felt like an earthquake. Demolition of the nearby public library had shaken the structure of the museum, and Gallery 33 was closed immediately that day to protect the artefacts.

It never re-opened. So the experiment begun over a quarter of a century before came to an end with something of a whimper.

What generic lessons can we learn from the birth and death of this flagship ethnographic gallery? What is its legacy and why should we bother to reflect on it now? First, despite the caveats above, Gallery 33 is a success story. A 26-year lifespan is longer than most permanent displays. It captured the ambitions of ethnographic display in a way few other competitors managed. It moved intellectually beyond the early multiculturalism of the Museum of Mankind. It helped redefine Birmingham's self-image through its celebration of Brummagem ware in both artefactual and human ways. It engaged visitors in ways that only its successor history galleries have been able to achieve. But as do all ethnographic displays, it foundered. Lubar suggests three reasons that museum collections die: they become stagnant, they are museums that nobody uses and their original curators abandon them (Lubar et al. 2017, 7). All three factors were involved in the demise of Gallery 33, and all were inevitable.

There are some failures specific to Gallery 33. Long-term financial and intellectual investment costs were not considered at the outset. Intellectual investment also included the outreach programme, with school and community events held at the small stage at the rear of the gallery. These events had a fitful and short life and, crucially, made few effective links with the various Afro-Caribbean and Asian communities in Birmingham. Nor was attention paid to how permanent the display was to be. Linked to this are more general issues. Futureproofing is often involved in museum display decisions but was absent from the display in Gallery 33. Images of Benazir Bhutto and Mikhail Gorbachev were small indicators that the display was stuck in a static ethnographic present. But, I would argue, there is no ethnographic present which avoids the issues of power, subjectivity, the status of artefacts, representation and facilitation. Ethnographic displays are complex artefacts anchored in their historic heritage. Furthermore, attempts to futureproof have two negative consequences: they drain any sense of liveliness from the display, and they suffer the 'Ames effect': they always remain anthropological boxes 'freezing' others into academic categories (Ames 1992, 140).

A larger problem beset Gallery 33 and it is one that affects all ethnographic displays. 'A meeting ground of cultures' derives from the concept of 'World Museums and World Cultures', the contemporary successor to earlier multicultural and anti-racist programmes. What the concept insists is that propinquity breeds respect. There is little evidence to either support or reject this claim. It may well be that museums are not the best place to promote such agendas unless further work is undertaken to ensure that some form of 'chemical reaction' can be effected, by bringing together similar social practices and beliefs in different societies to form new cultural 'compounds'. Ethnographic and other museums rely upon both 'resonance' and 'wonder' to engage their audience. Gallery 33's highlighting of commonalities across cultures is an exercise in evoking resonance ('the power of the displayed object to reach out beyond its formal boundaries to a larger world') but it also invokes the older sentiment of 'wonder' ('the power of the displayed object to stop the viewer in his or her tracks, to convey an arresting sense of uniqueness, to evoke an exalted attention') (Greenblatt 1991, 42). Wonder as a concept is tricky in an ethnographic setting and inevitably risks invoking its historic precedent, the Cabinet of Curiosities. This was implicitly what the members of the African Caribbean community recognised in Gallery 33's style of presentation and the heightened attention paid to the history of its artefacts and their collectors.

What has to be conceded is the courage and commitment that gave rise to Gallery 33. The 1990s was a turbulent period in Birmingham's history and there were loud voices demanding ethnic representation in new museum displays. Gallery 33 did not ignore these voices; indeed, the Advisory Group had a majority of members from the various communities in the city, but it steered a perilous path between specialist arguments in search of a common language. Initially the project worked well, but inevitably, the hubris of neglect blunted its message and led eventually to its demise. However, it made its mark on the history of British ethnographic display. We may call it a heroic failure, and failure is to be expected and embraced in any discussion of ethnographic display. We might go further and be tempted to concede that failure and loss provide the space for 'the emergence of new values, attachments and forms of significance' (DeSilvey and Harrison 2020, 3).

Perhaps two features specific to Gallery 33 stand out now for our reflection. The first, the failure of the interactive video, has important lessons as we embrace new technologies. But the second, the interrogative approach to imparting new knowledge that was so pervasive in the exhibition may, perhaps, suggest that this approach was a concealed version of the rhetorical question – making a point rather than expecting an answer. Does it still remain licit to introduce an agenda – however laudable its intent or content – without declaring this underlying objective? This is a painful question that museum ethnographers must constantly battle with. Perhaps we should be a little more open in declaring our missions. World Museums have still not solved the issues of cultural representation that Gallery 33 grappled with.[3]

Notes

1 *Collectors in the South Pacific*, 1989. LaserVision Interactive Videodisc installation produced by New Media Productions, London.
2 Chris Wingfield interviewed by Nick Stanley, 2016.
3 I am grateful for advice and comments from Adam Jaffer, Mick Orr and Chris Wingfield.

References

Akomfrah, J. 1986. *Handsworth Songs* [documentary film]. Black Audio Film Collective and Channel 4 TV.

Ames, M. 1992. *Cannibal Tours and Glass Boxes: The anthropology of museums*. Vancouver: UBC Press.

Appadurai, A. and N. Alexander. 2020. *Failure*. London: Polity Press.

Baxandall, M. 1991. 'Exhibiting intention: Some preconditions of the visual display of culturally purposeful objects'. In *Exhibiting Cultures: The poetics and politics of museum display*, edited by I. Karp and S. Levine, 33–41. Washington, DC: Smithsonian Institution.

Birmingham Gazette. 1931. 'Birmingham boon: Interesting addition to museum'. *Birmingham Gazette*, 5 May.

Birmingham Mail. 1931. 'South Sea Islands: Fascinating exhibition at art gallery. Mr. A. W. Wilkins gift'. *Birmingham Mail*, 4 May.

Birmingham Post. 1928. 'Branch art galleries and museums'. *Birmingham Post*, 3 May.

Birmingham Post. 1931. 'Native objects of the South Pacific: Large collection for the Birmingham Museum'. *Birmingham Post*, 5 May.

Bouquet, M. 2001. *Academic Anthropology and the Museum: Back to the future*. Oxford: Berghahn.

Bridgman, R. and A. Jaffer. 2017. 'Faith in Birmingham: A new voice for people of faith in the city'. *Journal of Museum Ethnography* 30: 104–21.

Burt, B. 2019. The *Museum of Mankind: Man and boy in the British Museum Ethnography Department*. Oxford: Berghahn.

Cannizzo, J. 1989. *Into the Heart of Africa*. Toronto: Royal Ontario Museum.

Delisse, C. 2020a. 'Against the mono-disciplinarity of ethnographic museums'. In *Across Anthropology: Troubling colonial legacies, museums and the curatorial*, edited by M. von Oswald and J. Tinius, 231–41. Leuven: Leuven University Press.

Delisse, C. 2020b. *The Metabolic Museum*. Berlin: Hatje Cantz Verlag.

DeSilvey, C. and R. Harrison. 2020. 'Anticipating loss: Rethinking endangerment in heritage futures'. *International Journal of Heritage Studies* 26 (1): 1–7.

Engels, F. 1886/1968. 'Ludwig Feuerbach and the end of classical German philosophy'. In *Karl Marx and Frederick Engels: Selected works*, 596–632. London: Lawrence Wishart.

Fennell, Catherine. 2023. 'Afterwords: Failed projects'. *Journal of the Royal Anthropological Institute* 29: 190–6.

Greenblatt, S. 1991. 'Resonance and wonder'. In *Exhibiting Cultures: The poetics and politics of museum display*, edited by I. Karp and S. Lavine, 42–56. Washington, DC: Smithsonian Institution.

Grimley, T. 1990. 'Treasure trove of global dimensions'. *Birmingham Post*, 9 October.

Holbraad, M. 2011. 'Can the thing speak?'. OAC Press Working Papers Series #7. Open Anthropology Cooperative Press. Accessed 27 September 2024. https://openanthcoop.net/2011/01/12/can-the-thing-speak/.

Honan, William H. 1990. 'Say goodbye to the stuffed elephants'. *New York Times Magazine*, 14 January, Section 6, 34.

Kahn, M. 1995. 'Heterotopic dissonance in museum representation of Pacific Island cultures'. *American Anthropologist* 97 (2): 324–38.

Karp, I. 1992. 'Audience, ownership and authority: Designing relations between museums and communities'. In *Exhibiting Cultures: The poetics and politics of museum display*, edited by I. Karp and S. Lavine, 136–57. Washington, DC: Smithsonian Institution.

Karp, I. and S. Lavine. 1993. 'Partners in crisis'. *Museum News*, May/June: 1–9.

Lubar, S., S. Rieppel, A. Daly and K. Duffy. 2017. 'Lost museums'. *Museum History Journal* 10 (1): 1–14.

Macdonald, S. 2020. 'Transforming the ethnographic: Anthropological articulations in museum heritage research'. In *Across Anthropology: Troubling colonial legacies, museums and the curatorial*, edited by M. von Oswald and J. Tinius, 49–62. Leuven: Leuven University Press.

Modest, W. 2020. 'Museums are investments in critical discourse'. In *Across Anthropology: Troubling colonial legacies, museums and the curatorial*, edited by M. von Oswald and J. Tinius, 65–74. Leuven: Leuven University Press.

O'Hanlon, M. 1993. *Paradise: Portraying the New Guinea Highlands*. London: British Museum Press.

Peirson Jones, J. 1987. 'Gallery 33 project: Initial design brief, September'. Birmingham: Birmingham Museum Archives.

Peirson Jones, J. 1990. 'Interactive video and the Gallery 33 project'. *Museum Development* June: 1–8.

Peirson Jones, J. 1991a. 'Interpreting culture: The colonial legacy and the community'. Paper read at Museum Ethnographers Group Conference, Birmingham Museum, 19 March 1991. Birmingham: Birmingham Museum Archives.

Peirson Jones, J. 1991b. '"Gallery 33: A Meeting Ground of Cultures": A new anthropology gallery in Birmingham'. *Birmingham Museum News* 49, January.

Peirson Jones, J. 1992. 'The colonial legacy and the community: the Gallery 33 project'. In *Museums and Communities: The politics of public culture*, edited by I. Karp, C. M. Mullen Kraemer and S. D. Lavine, 221–41. Washington, DC: Smithsonian Institution.

Peirson Jones, J. with P. McManus, P. Watson and S. Werner. 1993. 'Gallery 33: A visitor study'. Birmingham: Birmingham Museums and Art Gallery.

Peirson Jones, J. with N. Stanley. 2007. '"A meeting ground of cultures": Reflections on the genesis of "Gallery 33"'. *Journal of Museum Ethnography* 19: 70–4.

Phillips, R. 2003. 'Community collaboration in exhibitions: Towards a dialogic paradigm: Introduction'. In *Museums and Source Communities: A Routledge reader*, edited by L. Peers and A. Brown, 155–70. London: Routledge.

Riegel, H. 1996. 'Into the heart of irony: Ethnographic exhibitions and the politics of difference'. In *Theorising Museums: Representing identity and diversity in a changing world*, edited by S. Macdonald and G. Fyfe, 83–104. Oxford: Wiley-Blackwell.

Rodman, M. C. 1993. 'A critique of "place" through Field Museum's Pacific exhibits'. *The Contemporary Pacific* 5 (2): 243–74.

Rubio, F. D. 2016. 'On the discrepancy between objects and things: An ecological approach'. *Journal of Material Culture* 2 (1): 59–86.

Shelton, A. 2009. 'The public sphere as wilderness: Le Musée du quai Branly'. *Museum Anthropology* 32 (1): 1–15.

Shelton, A. 2013. 'Critical museology: A manifesto'. *Museum Worlds: Advances in Research* 1: 7–23.

Simpson, M. 1992. 'A meeting ground of cultures: A review of Gallery 33, Birmingham Museum and Art Gallery'. *The Social History Curators Journal* 19: 72–7.

Sunday Mercury (Birmingham). 1931. 'The Headsman's sword: Gruesome relics for Birmingham'. *Sunday Mercury*, 28 June.

Terrell, J. 1991. 'Disneyland and the future of museum anthropology'. *American Anthropologist* 93 (1): 149–53.

Thomas, N. 2015. 'A critique of the natural artefact: Anthropology, art and museology'. Art History Lecture Series 13. Wellington: Victoria University of Wellington.

Thompson, N. 1989. 'The ROM apologies for 1989'. *Global News* (Toronto), 10 November.

Wingfield, C. 2006. '(Before and) after Gallery 33: Fifteen years on at Birmingham Museum and Art Gallery'. *Journal of Museum Ethnography* 18: 49–62.

Part II
Current practices

4
'Tervuren remains a place of false memories': on the impossibility of an epistemological rupture at the Royal Museum for Central Africa (Belgium)

An interview with Boris Wastiau

Boris Wastiau and Arnaud Lismond-Mertes

Translated from the original French by Boris Wastiau[1]

Boris Wastiau was the curator of the exhibition *ExItCongoMuseum: A century of art with/without papers*, presented in Tervuren in 2000. In the words of Philippe Marechal, the Royal Museum for Central Africa's *ad interim* director at the time, the exhibition was considered 'an impulse for a more global programme: the renovation and updating of the exhibition spaces' (Wastiau 2000a, 5). Advocating a 'critical museology' (following Anthony Shelton), his *ExItCongoMuseum* was an open, reasoned and structured challenge to what had been practised in Tervuren until then. It provoked contradictory reactions within the museum itself (Corbey 2001; Capenberghs 2001). Boris Wastiau wrote the catalogue for this exhibition (Wastiau 2000a), and in subsequent publications returned to the RMCA's museology and the way the objects were brought in and presented (Wastiau 2000b; 2001; 2002; 2003; 2004; 2005; 2008a; 2017). An anthropologist and a curator, he is also the author of a number of books (Wastiau 2006; 2008b; 2016; 2018). At the time of the original interview, he was director of the Geneva Museum of Ethnography (MEG) and professor at the University of Geneva. He had returned to Brussels to discover the new permanent exhibition at the RMCA. We met him as he wrapped up two days spent visiting the museum. It was an opportunity to benefit from the view of a man who has been fully involved in the international debates on the future of ethnographic museums and their museology for more than 20 years. He also combines an intimate knowledge of the Tervuren museum (he worked there for 11 years, from 1996 to 2007) with a 'deep attachment' to the institution from a critical, sometimes radical, distance.

Arnaud Lismond-Mertes: What are your first critical impressions of this new permanent exhibition and in particular of its stated 'decolonial' ambition?

Boris Wastiau: As the director of a museum of ethnology, I hear your question. It is my duty to take part in debates about museums that are open to society, especially when I am invited to do so. But I come to this interview in a state of mind, divided between a critical point of view, albeit a well-founded one, between my feelings as a Belgian and a European, as an advocate of an open and cosmopolitan society, sensitive to the impact that such a museum display might have, and lastly between my feelings of respect and loyalty to the members of the RMCA, where I used to work, and to the director, now my counterpart.

It was with great interest that I rediscovered this museum in its new configuration, having come to try to approach it with fresh eyes, with as few preconceptions as possible, forgetting my expectations as much as I possibly could. My first impressions of the new permanent exhibition were quite positive. I found all the amazing diversity of the museum's unequalled collections. I was delighted to see many exceptional art pieces that I had not seen for more than 10 years. All of this was presented with a clear desire to bring in contemporary perspectives, using multimedia and a large number of screens that gave a voice to the Congolese people and to people from the diaspora. This is something that had never been done in the permanent exhibition of the Tervuren Museum. But after my visits and after reading the museum guide, I wondered if we had not returned to 'the Congo Museum'.

The scenography of the exhibition did not seem to me to be very inviting or very coherent. It has to be said that the museum does not provide the public with a clearly defined route to follow through the exhibition. This may have its advantages and offer the pleasures of serendipity. But it also has the disadvantage of not inviting the public to think in an organised way. After passing through the underground entrance corridor and walking along an impressive dugout canoe (I did not understand why it was there), you come to a small section which introduces the museum's website, research services and collections. It is good to start there, with the question of the collections' provenance. But the way in which the museum talks about this seems to me to be extremely flimsy in relation to what has now been learned and published about the violence of colonial collecting and colonialism in general. The museum points out that since colonisation began, administrators, traders and missionaries have 'collected' objects – using that perennial euphemism for appropriation – but insists on there being a scientific approach to the process. However, the motivation was not primarily scientific for 90 to 95 per cent of the so-called 'ethnographic collections'. That was a key fact presented in my exhibition in 2000. The question of the acquisition of these objects, which has been the subject of much public debate in the context of restitution claims, should by now be the subject of a detailed examination. Neither in this part of the museum's presentation nor systematically in all parts of its permanent exhibition, including the natural sciences, does the museum really address this issue.

However, the Tervuren Museum, like the Ethnographic Museum in Geneva, which I direct, is almost entirely made up of colonial collections that followed the same routes and were subject to the same processes of appropriation and display until sometime late in the twentieth century. It therefore seems important to me to show visitors the asymmetry of exchange in the colonial world through the objects on display, and to explain that these objects and specimens were 'collected' in the same extractivist spirit that prevailed in mining, the exploitation of timber, rubber or other 'available resources'. This unilateral extractivist exploitation of the Congo is not at all evident in the presentation chosen by the museum. Which objects, if any, were acquired legally? Were the objects bought from people who had the right to sell them? Was consent always given voluntarily? Were the terms of the transactions fair, given the power relations? Were the objects cultural objects of religious, political or genealogical significance to the original populations? Furthermore, the objects are presented without the Congolese people's own words about them, without research into who made them, how they came to be in the museum, etc. There is no shortage of information on this in the institution's archives, though!

Beyond this essential aspect of the presentation of the objects in the collection, I was most interested to see how the 'decolonial' question was treated in this new permanent exhibition. In my opinion, this question is not only about history, but also about the current consequences of the colonial period, the resurgence of colonial-type relationships, for example in terms of racial discrimination. I was left wanting in this respect ... In fact, although colonisation and decolonisation are dealt with in a relatively factual way in the museum's historical section, this section remains a minor part of the museum. Colonialism as a subject matter is not at all mainstreamed in the overall presentation. All the collections, all the scientific disciplines and all the ways in which they are presented that are rooted in the colonial period need to be questioned. In the former colonial and ethnographic museums, whether in Tervuren or Geneva, it is therefore all the scientific disciplines and the museum as a whole that should be subjected to decolonial reflection. Of course, this is not something that can be done overnight, but the modernisation of the museum in Tervuren has officially been under way for more than twenty years …

The museum's section on 'Rituals and ceremonies' through the stages of life is emblematic in this respect. It provides information that is factually correct. We see Congolese people on screens talking about all sorts of institutions (marriage, education, initiation). But the very structure of this room remains the colonial structure of ethnographic museums as they existed in the 1960s, which at that time 'benevolently' showed the functioning of 'indigenous' societies 'from birth to the afterlife', through their fertility, marriage and funeral rites, and so on. Nothing exhibited here is wrong in itself, but it is clear to me that this way of presenting the museum's heritage does not respond to current social and political issues. Today, the reality of the world experienced by most of our contemporaries is no longer based on presumed singular ethnic identities but has become 'translocal'. Individuals mobilise multiple identity processes, different from one generation to the next, different over the course of a lifetime, and they are mobile. Africa and the Congo are not only a continent and a country, but they are also cultural, historical and social spaces in the world, especially in Europe and in the Americas. Today, one can be a citizen of Kinshasa, or 'Kinois', while living in Dubai or Brussels. One is 'Kinois' because one identifies with a particular culture, but it is no longer simply a question of place. This is something that ethnographic museums, which have always sought to classify and 'fix' people (in a region, a territory, a language, a culture and with a particular identity), still find very difficult to capture.

AL-M: Tervuren remains the 'museum of the Others' …

BW: In a way, yes … As for the section of the museum devoted to 'African art', it is in the register of what was done at the end of the 1980s, when the objects (which over time had been presented successively as trophies of conquest, then as exotic, ethnographic, scientific and artistic objects) were presented as 'masterpieces'. The curator's choice is fantastic and pays tribute to the mastery of Congolese sculptors of the past. There is no doubt about it! Some pieces had been on display for many years, while others were less well known. Of course, it seems to me perfectly legitimate to share with the public the appreciation of the aesthetic qualities of the objects, even if many of the selection criteria are not deconstructed and it remains difficult for a non-specialist audience to approach them in anything but a completely subjective way … and it is a bit thin and reductive as an approach in the context of a museum, where so many expectations and questions are focused. And then, above all, one clearly feels that there is no articulation with the rest of the exhibitions presented in the museum and certainly not with the decolonial ambition of the institution. There are questions in the public debate about demands for the return of objects, such as a famous Kongo sculpture on display in this section. But the uninformed visitor will not know this.

With regard to the sections on natural resources and biodiversity, I think it is very good that the museum approaches these issues from a contemporary point of view: deforestation, ecological problems, etc. But once again, it seems to me that these issues would have benefited from being placed in a historical perspective and, in particular, in relation to the lasting consequences of colonisation and the period of independence. Today's economic, ecological, political, and human problems are the direct result of colonialism and post-colonialism. This is so obvious.

Regarding the history gallery, it seems to me that it does not take sufficient account of the fact that the general public has a rather poor knowledge of what colonialism was and tends to reduce it to the relationship between the colonising state and its colonies. In our case, Belgium on the one hand and Congo, Rwanda and Burundi on the other. Since the end of the nineteenth century, colonialism had been an offshoot of globalisation and extremely aggressive European and North American imperialism. It was not 'the Belgians' who went to colonise 'the Congolese', but very powerful economic interests that used all the means at their disposal (scientific, technologic, diplomatic, state and other) to exploit colonies through strategic arrangements. This is the global application of an extractivist logic of appropriation and exploitation of 'natural resources' that started in the sixteenth century, which also considers human beings as such and as likely to be squeezed like lemons. The International African Association and the Congo Free State (CFS), which achieved the conquest and domination of the Congo, were just that. It was not an initiative of Leopold II alone. It was an enterprise in which, until very late, the Belgians did not play as large a role as one might have imagined. Among others, American, Swiss, French, Scandinavian and British interests were also involved. The Berlin International Conference of 1884–5, which recognised the

Congo Free State as a sovereign state, was attended and signed not only by Belgium and Germany, but also by the United States, Great Britain, France, Portugal, the Ottoman Empire, Sweden-Norway and many other countries that called themselves 'civilised' at the time. The Conference set some rules of colonisation: the establishment of borders, free trade of some of the largest rivers, free establishment of (Christian) missionaries of all denominations regardless of origin. The colonisation of the Congo was therefore an international affair from the outset. This is something the museum has never addressed and, for reasons I cannot fathom, seems unwilling to do so. Could it be that it would amount to acknowledging that the Belgians and the Belgian state were never the masters of the game even after the end of the Congo Free State in 1908? The explorer Stanley was an American. Among the architects of the legal structure of the Congo Free State was a prominent Swiss lawyer, co-founder of the International Red Cross, Gustave Moynier, etc.! The Congo Free State operated by granting huge concessions of territory to large Belgian companies, but also to American, French, English and Scandinavian companies. Nokia's fortune began in the Congo. J. P. Morgan, Thomas Fortune Ryan, John D. Rockefeller and Daniel Guggenheim were big investors in the exploitation of mines and forests in the Congo Free State. The same is true of the Lever brothers in the palm oil industry, who laid the foundations for Unilever, etc. This global dynamic is clearly not explained at all in the exhibition. It would not excuse anything, of course, but it would enlighten those naive minds that still wonder 'how such a small country could colonise such a large territory!'

As for the treatment of the question of national independence in the gallery dedicated to 'history', it also seems to me to be treated in an excessively superficial manner. It is presented as a purely Belgian-Congolese affair and all its political aspects, both Congolese, Belgian and international, are completely ignored. However, if independence was finally achieved, it was because there were struggles led by the Congolese themselves, because there were political oppositions in Belgium too, but there was also an international context, with other struggles and tensions, both in Africa and on a global geostrategic level, in which all these events took place. All this is erased in the exhibition, and the Congo is presented almost as an island, with Belgium as its only horizon. But it is interesting to try to explain why independence came about, why it came about at that time, and what was the sequence of independence of neighbouring or more distant but influential countries, and so on. This whole history is part of an international context that plays a major role and concerns international issues related to the control of key strategic subsoil resources, such as copper, or the uranium from the Shinkolobwe mines, which enabled the Americans with their 'Manhattan Project' to carry out the two atomic bombings of Hiroshima and Nagasaki in 1945. While these hastened the end of the Second World War, they also marked the beginning of the Cold War. That the 'Cold War' between East and West led to the most insidious and violent interventionism on the African continent, facilitating the rise of dictators and sustaining civil wars for several decades, is clear to anyone who has read more than one book on contemporary history. The Tervuren exhibition refers to the 'chaotic beginnings' of independence in 1960–4 and goes on to say: 'The post-colonial history of Burundi, Congo and Rwanda is dominated by complex, tragic and controversial conflicts' (p. 95 of the guide), but it does not set this context or attempt to explain it in any way. It seems to me that the museum does not give visitors enough information to enable them to understand the sequence of events and the driving forces behind them. Who were the interests and powers that pulled the strings and kept the leaders in place? Who prevented African states from taking full control of their natural resources? These seem to me to be the fundamental questions. Even my children, who are not particularly interested in the subject, know that one of the reasons for the wars in eastern Congo today is to control the production and export of coltan, a mineral used in the manufacture of their mobile phones … The era of 'colonisation' ended only to give way to neo-colonialism, which has been even more profitable, both financially and industrially. Today, violent relations linked to resource extraction take place in a different context, but they continue with these historical roots. The exhibition could shed light on this period and the world today by showing these continuities …

I must admit one thing. What completely 'finished me' was the section of the exhibition dedicated to the question of the exploitation of wild rubber as a 'natural resource' (section 'The resource paradox' in the museum guide, Royal Museum for Central Africa 2019, 105). It is presented in a short paragraph in a way that is completely detached from the question of the infamous 'severed hands' of the nineteenth century, mentioned briefly in the historical section, but without any new research having apparently been carried out in this field in the last 20 years, despite repeated promises by the museum to do so. Fortunately, the museum guide is more explicit on pages 84 and 85. But just opposite the showcases containing some samples of wild and processed rubber and a magnificent yet very oddly displayed ivory bust of Leopold II, a text panel (or page 104 of the guide) states the following in bold print, perhaps to give us hope: 'The region … remains economically attractive. This is despite the political instability … The wealth of resources does not prevent Central Africa from being relatively poor – the paradox of abundance. Good governance focused on sustainability would greatly benefit the well-being of the region.' No comment!

AL-M: You mention the different sections of the museum, but how do you read the intention and the overall narrative of the permanent exhibition? The museum is not very explicit on this point …

BW: For me, the theme of the exhibition organised by the museum is 'Belgium's Congo 60 years after independence'. Basically, it's about explaining to the public 'what the Congo is, what its resources are and what its current challenges are', 'that the colonial period was "difficult", but that we have a common history and common interests in the face of future challenges', 'that the diaspora is welcome to collaborate at the museum', 'that we do scientific research and that sharing this knowledge and culture will benefit everyone', and so on. A host of common-sense good intentions. That is the way I see things.

AL-M: In a recent article, you wrote that today ethnological museums 'have not only lost their original colonial function but also their later cultural-scientific function of representing the world' (Wastiau 2017). What do you mean by this?

BW: The idea I was expressing was that the claim of ethnological museums to fulfil a purely 'representational function' is completely outdated. This is one of the reasons why more than half of the so-called ethnographic or ethnological museums in Europe have changed their name in the past 25 years. They were colonial archaisms. No one can claim to represent 'the Other' univocally and unilaterally any longer as in a not-so-distant past: 'here are the pygmies', 'here is how the Eskimos live', 'here is the realm of the Boschimans', etc. Ethnography is dead. Contemporary museums work around transverse questions, dynamics and problems. They openly question their disciplinary field. They admit, acknowledge and warn their public about the subjectivity and partiality of opinions that are presented. This is also the decolonisation of museums! The contemporary museum retains a character of 'authority', naturally, but it seeks to share it, to place itself in the register of intersubjectivity and co-construction … In any case, it does not imply to the visitor that what the museum presents is an absolute and definitive truth, rather, that our knowledge is in constant development. The exercise of the power of representation that museums have is a political act that affects thousands or hundreds of thousands of people during their visit every year. It is therefore a responsibility that we must be aware of and share. When you are designing an exhibition, you have to ask yourself: Who are you going to invite into the process? With which voices do we want to share this power of 'representation'? Which audience are we talking to, to say what? What are the sensitivities, needs and aspirations of this audience, and what questions do they have?

AL-M: From the introductory room to the main exhibition galleries, there is a very strong affirmation of the 'scientific' character of the museum (its scientific publications are exhibited, its research projects presented, etc.). Then the museum seems to give visitors a lot of answers while inviting them to ask themselves few questions. Doesn't this reflect the very flaw in the claim to 'objective' representation that you denounce?

BW: You have put your finger on issues that arise for this museum ... as well as for others. The ethical obligation of museums exhibiting colonial collections to systematically indicate their provenance was the subject of the article I wrote and to which you referred. This was done for the first time in 2000 in Tervuren in the *ExItCongoMuseum* exhibition, where each object was systematically accompanied by a label stating its provenance, that is, the function of the collector and the reason for collecting it, the manner of acquisition, the exact place and dates whenever available, and of course the name of the creator, owner and user. This practice did not last long! Recently, in an article published in the *Oxford Handbook of Public History* (Wastiau 2017), I wrote that collection curators, as 'public historians', have an obligation to reveal all the sensitive aspects of the collections in their care, to bring together disparate elements such as archives, objects, photographs and biographies, and to demonstrate the historical relevance and value of the collections. In this text I have compared the practice in Tervuren before the renovation with the presentation of the African collections – spoils of war! – which have always been displayed in the Royal Military Museum (in the Parc du Cinquantenaire, Brussels). Incidentally, I am surprised that, in the current context, no one has revisited these presentations of those trophies in the Military Museum, which date, if not from the beginning of the twentieth century, at least from the interbellum period. Indeed, it seems essential to me that museums should give an account of how the objects in the colonial collections they display were assembled, and of the actions and decisions of the people to whom these objects belonged. Ironically, this is most evident in the Military Museum, where the object labels are 'in their colonial juice' and disturbingly honest.

For example: 'Tunic of a dervish killed in Redjaf. 1897. Gift of Commandant Laplume'. You can understand that if every object were presented in this way, it would change the general tone!

In *ExItCongoMuseum*, only three artists, carvers, were identified for 125 masks and sculptures. All the other objects, 122 of them, had a label mentioning 'unknown artist'. This is one of the things that most shocked the establishment. Traditionally, so to speak, 'ethnographic museums' did not want to know or mention the names of the people who made these objects, the circumstances of their collection, etc. We are talking about an object that came from a chief, but which chief? He had a name, perhaps even an opinion! We talk about an ethnologist's collection, but this ethnologist worked with people in the field. Who were his respondents? Finding these actors and bringing them out is, in my opinion, an integral part of the decolonisation of ethnological museums. It is going to be a painstaking task, but we must have the ambition to make known the actions, gestures and words of all the people who had agency in the colonial encounter, in all fields. Today, the Tervuren museum only obliquely mentions the actions of a few people, such as Lumumba or Habiyarimana, who are perhaps best remembered for the indignation they caused in Belgium. That is not much! It is a bit thin! Again ... What about all those who have been forcibly silenced? Was it not the right time to bring them back to light and make them known? Even the leading Congolese political figures: what were their voices, their messages? The museum does not really give the impression that it has any intention of talking about them.

AL-M: The museum presents the Great Dugout Canoe without giving the point of view of the Congolese on it but relating it to its construction for use by Leopold III in 1957 and mentioning that it does not know whether it was 'offered' to the former sovereign by the local population or whether its production was the object of an order from the colonial administration. If we follow your logic, shouldn't the museum have sent a researcher to the village where the canoe was built, in order to collect the point of view of the local actors or their memory, and then exhibit the results of this research in relation to the object itself?

BW: I think I have said enough and that you have understood me well. The museum would benefit from more research into the origins of its collections and the objects it exhibits. It seems to me that it should also start a work of co-construction of the exhibitions with the populations of origin and the other stakeholders. When I left the museum in 2007, we were not yet talking about the co-construction of knowledge, but

I left a few modest contributions on the provenance of its collections. Since then, there has been no systematic and ambitious work on the collection. On the contrary, the number of curators has been drastically reduced, from four to one. In 2015 Maarteen Couttenier, who now works at the RMCA, published an excellent article in a Dutch scientific journal on a nkisi nkonde statue currently on display in the 'Art without equal' exhibition gallery (Couttenier 2015). In this case, the museum has all the information it needs to talk about the fateful history of this object and the events it was involved in, as well as recent fieldwork data. But it will not do it. Why? Many other examples could be given. For instance, the museum has a definitely unique collection of religious art collected in 1884 by Lieutenant Emile Storms: dozens of objects taken by force from Tabwa chiefs in eastern Congo.

These objects are part of an 'infernal' affair which has long been known to specialists and which was recently brought to the attention of the general audience by the journalist Michel Bouffioux in 2018. But in the history section of the museum, where some of these objects are displayed, it says only: 'In 1884, Storms launched a bloody expedition against Lusinga. The chief and fifty of his men lost their lives … The ancestral statue representing Lusinga was taken as war booty by the Belgian officer. He also brought back the skull of Lusinga which is now in the Royal Belgian Institute of Natural Sciences.' The uninformed visitor who sees this and does not know who the Tabwa were or who Storms was is led to think: 'Well, we were told it was violent! There was a brawl and fifty dead …'. That narrative doesn't say what is behind it. The point is not just that people died, but that an entire royal Tabwa family (Lusinga and his cousins) were deliberately murdered, beheaded and their heads meticulously flayed and cleaned because, in the words of Storms in his notebooks, 'they would look good in the museum with a label on them'. And that is what he did. Today the museum holds the ritual religious objects that Captain Storms confiscated from the Tabwa as war trophies after the killings, as well as Storms' notebooks telling his glorious story and the soldier's memorabilia. The skulls were deposited decades ago at the Royal Belgian Institute of Natural Sciences in Brussels. Thanks to the abundance of historical data and material evidence about this particular story, it would be very easy to exhibit in a compelling manner and would say a lot about these objects and the grim reality of colonial conquest. Could it just be an unrepresentative, bloody event? No, it could not. You can find other cases *ad nauseam* to illustrate the numerous facets of the extreme violence of colonialism in Central Africa.

AL-M: In your writings on the old exhibition in the Tervuren Museum, you have repeatedly described it as a 'place of false memories'. Do you maintain this description for the new exhibition?

BW: Firstly, it can still be said that the AfricaMuseum in Tervuren remains a place of false memories, in the sense that the permanent exhibition explicitly purports to present the shared history and memories of the Belgians and the Congolese. So far, I have given you a few examples of how history has been truncated and erased there. But memories? What and whose memories? There has been no serious attempt to confront the memories of individuals from a wide range of social positions throughout the twentieth century. Not even the historical violence of race relations is problematised and thoroughly considered in the museum's presentation by means of memorial data. Secondly, it seems that what the museum presents is an institutional and political perspective that is simultaneously a defence and an apology for Belgium's colonial past. It is as if the museum were in the process of the construction of a schizophrenic institutional memory. I do not know how this construction was born and how it has been implemented, but it does not seem to me to be an expression of a real, shared memory between 'the Belgians' and 'the Congolese'. Yes, there are Africans and Afropeans (their word) who express themselves here and there in the museum through a large number of display screens. But during my visit, I did not have the feeling that the Congolese and the Congolese diaspora had appropriated the museum and that they were really 'at home' there, expressing 'shared memories'. And then again, you see, the *AfricaMuseum* only talks about itself, of 'the Congo and the Congolese', but not about Africa or Africa in the world and in history. Perhaps is it because the museum has a lot of potential that it is causing so much controversy today, because there are very strong energies that are clashing in this place, because something is happening, because it has things to say to us that will need to come out? Maybe because of that something new will eventually come out of it.

Note

1 This translated interview is based on one originally published in French in the magazine *Ensemble!* (No. 99, May 2019), available at http://www.asbl-csce.be/journal/Ensemble99dossier39 (Accessed 27 August 2024). Arnaud Lismond-Mertes, of Collective Solidarity Against Exclusion (Belgium), asked anthropologist and Africa specialist Boris Wastiau to give his first impressions of the new permanent exhibition of the AfricaMuseum (Tervuren, Belgium – aka the Royal Museum for Central Africa) after years of renovation and a declared process of 'decolonisation'.

References

Capenberghs, J. 2001. 'Re-turning the shadow: Museological reflections on the Royal Museum for Central Africa in Tervuren, Belgium'. ICOM Conference, Barcelona, 2–4 July, 2001. Accessed 28 August 2024. https://icme.mini.icom.museum/wp-content/uploads/sites/16/2019/01/ICME_2001_capenberghs.pdf.

Corbey, R. 2001. 'ExItCongoMuseum: The travels of Congolese art'. *Anthropology Today* 17 (3): 26–8.

Couttenier, M. 2015. 'EO.0.0.7943'. *BMGN-Low Countries Historical Review* 133 (2): 79–90.

Royal Museum for Central Africa. 2019. *Royal Museum for Central Africa: Guidebook*. Kontich/Tervuren: BAI/RMCA.

Wastiau, B. 2000a. *Congo-Tervuren, aller-retour: Le transfert de pièces ethnographiques du Musée royal de l'Afrique centrale à l'Institut des Musées nationaux du Zaïre* [Congo-Tervuren, round trip: The transfer of ethnographic pieces from the Royal Museum for Central Africa to the Institute of National Museums of Zaire]. Tervuren: Musée royal de l'Afrique centrale.

Wastiau, B. 2000b. *Mahamba: The transforming arts of spirit possession among the Luvale-speaking people of the Upper Zambezi*. Fribourg: Fribourg University Press.

Wastiau, B. 2001. 'The universe of Miss Jeanne Walschot, collector and dealer of "Congoliana", 1896–1977'. In *Collectors: Individuals and institutions*, edited by Anthony Shelton, 223–38. London and Coimbra: Horniman Museum and Museu Antropologico da Universidade de Coimbra.

Wastiau, B. 2002. 'La reconversion du musée glouton [The reconversion of the gluttonous museum]'. In *Le musée cannibale*, edited by Marc-Olivier Gonseth and Jacques Hainard, 85–109. Neuchâtel: Musée d'Ethnographie de Neuchâtel.

Wastiau, B. 2003. 'Les plaques sensibles de la mémoire "ethnographique": Congo belge, 1890–1930 [The sensitive plates of 'ethnographic' memory: Belgian Congo, 1890–1930]'. In *X: Spéculations sur l'imaginaire et l'interdit*, edited by GHK, 239–65. Neuchâtel: Neuchâtel Museum of Ethnography.

Wastiau, B. 2004. 'From Mpala to Tervuren: The ritual life of two Congolese masterpieces of the Royal Museum for Central Africa, 1880–2000'. In *Science, Magic and Religion: The ritual processes of museum magic*, edited by Mary Bouquet and Nuno Porto, 95–115. Oxford: Berghahn.

Wastiau, B. 2005. 'L'héritage colonial dans la conception de l'art congolais [Colonial heritage in the conception of Congolese art]'. In *Les rites et les arts*, edited by Amos Fergombe, 18–29. Valenciennes: Université de Valenciennes.

Wastiau, B. 2006. *Chokwe*. Milan: 5 Continents.

Wastiau, B. 2008a. 'Managing the African art collection history: Acquisitions, documentation and exhibiting practices at the Royal Museum for Central Africa, 1897–2005'. In *Museia 5: Museums and collection management issues*, 54–66. Thessaloniki: Aristotle University and University of Macedonia.

Wastiau, B. 2008b. *Medusa: The African sculpture of enchantment*. Geneva and Milan: MEG and 5 Continents.

Wastiau, B. 2016. *Amazonie: Le chamane et la pensée de la forêt* [Amazonia: The shaman and the thought of the forest]. Geneva and Paris: MEG and Somogy.

Wastiau, B. 2017. 'The legacy of collecting: Objects, images and people from colonial Congo'. In *The Oxford Handbook of Public History*, edited by Jim Gardner and Paula Hamilton, 460–78. Oxford: Oxford University Press.

Wastiau, B. 2018. *Afrique: Les religions de l'extase* [Africa: The religions of ecstasy]. Geneva and Paris: MEG and Somogy.

5

Ethnographic collections at Queensland Museum: histories and politics of exhibiting in a settler-colony

Chantal Knowles

In August 2016, Queensland Museum closed its first and only dedicated Aboriginal and Torres Strait Islander cultural gallery, *Dandiiri Maiwar*, after just 11 years. The closing ceremony speeches emphasised a temporary loss but a brighter future. A new and larger exhibition was planned to take centre stage on the museum's entrance level, with an expected opening within two years. The closure was necessitated by the need to make room for *Wild State*, the new natural history gallery, which opened in November 2016. As a result, the number of Indigenous objects on permanent display plummeted to less than 100 items overnight.

Queensland, the sole state which is the ancestral place of Australia's two Indigenous communities – Aboriginal and Torres Strait Islander Peoples – is also the main home to the Australian South Sea Islander population, descendants of western Pacific labourers indentured in the late nineteenth century. The region maintains strong and active cultural ties with Papua New Guinea through the Torres Strait islands. However, despite this rich cultural tapestry, these Indigenous cultures remain largely invisible on the museum's floor. Almost a decade later, a new Aboriginal and Torres Strait Islander gallery has yet to be opened.

Queensland's colonial history and evolving identity were intricately linked to the narratives presented by the museum in its early years. These narratives were shaped by the economic, social and political conditions of the time. As a state institution, the museum prioritised natural science collections and associated staff, recognising their potential to contribute to economic development and garner worldwide recognition for the region's unique biodiversity. In contrast, the acquisition of cultural materials followed a sporadic, incidental, and reactive pattern.

The events of 2016 are one of a series of acts that punctuate 150 years of the exhibition and interpretation of the 'ethnographic collections'. Here I review decades of collecting, collections management, display practices and museum governance to gain a deeper understanding of the interplay between collections, the museum's vision and purpose and how these aspects are manifest in public displays. The goal is to understand better significant moments like these, which contextualise the museum within both its evolving civic landscape and its often ambiguous relationship with its cultural collections. The museum's collections and displays are reviewed in relation to local, national and international legislation and politics, which framed and reframed the collections and attitudes to their display and care.

Queensland Museum was conceived and founded in relation to Queensland and Australia's position in the British Empire. Queensland was part of the empire and was involved in acquiring new colonies (British New Guinea in 1883) and international conflicts, on behalf of Britain, including the Boer, First and Second World Wars. At home in Australia, local politics continually defined domestic relationships with Aboriginal and Torres Strait Islander Peoples and Australian South Sea Islanders.

In the immediate post-war era, the formation of the United Nations and UNESCO articulated global thinking around human rights and Indigenous rights, including rights to education, property and collections. At the same time, many British colonies navigated towards independent nation-states. These global events and debates influenced museum practice and consideration of cultural collections. In Australia, the Pigott Report, *Museums in Australia 1975*, followed by the 1978 UNESCO *Regional Seminar on the Role of Museums in Preserving Indigenous Cultures*, held in Adelaide, which Queensland Museum staff did not attend, set out the issues facing museums and led to increased Indigenous consultation, followed by the proactive recruitment of Indigenous staff in museums. Guidelines followed, written by the professional groups and senior museum staff. These included *Previous Possessions, New Obligations* and *Continuous Cultures, Ongoing Responsibilities* (Museums Australia Inc. 1993; 2005) and most recently, the Australian Museum and Galleries Association's *Indigenous Roadmap* (Janke 2018).

Mirroring global politics and national and state legislative changes, the cultural collections underwent several identity shifts. Initially referred to as 'curios', they were later categorised as ethnology, and this term lives on in the titles of various associated object registers. Over time, the renaming of the discipline in relation to curatorial care has attempted to adjust museum relationships with the collections. From the 1960s, curatorial sections and collections have been named Anthropology, Archaeology, Kastom collections and First Nations, among others.

In this chapter, the cultural collections are used as both educational and political tools in exploring the history of Queensland Museum in relation to the history of the colony and the later state and its relationship with Indigenous people in Australia and the Pacific.

World's fairs and ethnography in the early days

Queensland Museum was founded in 1862, three years after the colony's separation from New South Wales. The first indentured South Sea Islanders were brought to Queensland to work the sugar cane fields the following year. Brisbane grew with purpose, and essential institutions, including the Parliament and Supreme Court, were convened in the former military and convict barracks as the city embarked on a building programme to meet the colony's needs.

The museum supported the growth of this nascent capital city through its close association with Queensland's contributions to world's fairs (McKay 1997; 2004). Queensland won awards for its display designs. Among its most memorable exhibits were a mercury fountain installed at the Greater Britain Exhibition in London in 1899 and a series of gold-painted Corinthian columns representing individual goldfield output (McKay 2004, 47–9). These popular fairs were supplemented by publications and photographic displays which advertised opportunities for investors and migrants. Museum staff contributed material and interpretation and sometimes received exhibits and casework in return. This was not just a sharing of resources; there was also shared sentiment, with the museum mimicking the Fairs by 'showing off' economic opportunities to locals and new arrivals. Queensland entered its first international exhibition in London in 1862, where it won attention as the youngest of all British Colonial possessions. The colony's government promoted a 'land of opportunity', with rich natural resources ripe for exploitation, including timber, possums for their skins and furs, dugong for oil and vast tracts of land suitable for pastoral and agricultural leases. By the London Annual International Exhibition of 1871, Queensland began a more targeted campaign to advertise economic opportunities for investors and immigrants. For this display, Richard Daintree, the

Government surveyor for Northern Queensland, advocated the inclusion of mineral specimens and also his photographs, to better represent the colony. The government appointed him a special commissioner, and the Queensland Annexe at the London International Exhibition was installed and accompanied by a published illustrated guide *Queensland, Australia* (Daintree 1873).

The museum-style displays, visible in the 1872 annexe, comprised desktop cases of specimens against a backdrop of photographic illustrations on the wall. Queensland's First Nations population were presented as 'set-dressing' through the fanned display of weapons and shields flanked by two of Daintree's hand-tinted large format portrait photographs of Aboriginal men. These displays toured to Vienna for the Universal Exposition of 1873, then to Philadelphia for the United States Centennial Exhibition. Queensland contributed to the Paris exhibition (1878) and Daintree's successor, Henry Ashwell, maintained the display style.

In 1879–80, at the Sydney International Exhibition, engaging taxidermy tableaux were added, including one entitled *No Laughing Matter* (featuring a carpet snake attacked by three kookaburras) by Anthony Alder. An Ethnological Court was placed above Queensland's exhibit to display South Pacific (principally New Guinea) collections sourced from Queensland Museum's collection with the addition of a small number of Aboriginal and Torres Strait Islander artefacts.

Queensland Museum, founded on the principles of learning and civilisation through the Queensland Philosophical Society, was heavily influenced by the economic, educational and entertainment factors that could lead to its success. In these early days, the museum provided all these functions for the settler population and became an extremely popular venue for both visitors to the city and locals (see Bennett 2005). Visitor numbers as a proportion of the population were high. In 1886, a census year, the Brisbane population was 51,689 and visitor numbers reached 106,907 for the year – equivalent to a third of the colony's population. On Sundays, the museum was particularly busy: 'It is really a pleasure to see so respectable and orderly a throng, and one so thoroughly appreciating the innocent recreation provided for it' (De Vis quoted in Bather 1894, 229). This 'innocent recreation' included the Darwinian tropes of 'dying races' and displays of ancestral remains.

Despite episodic acknowledgements of the value of technology, industry and social and cultural objects in the annual reports, cultural objects were rarely described in detail nor defined as a discrete collection. With the few resources at the museum's disposal, natural sciences were prioritised. In addition to his duties as museum custodian, Karl Theodor Staiger worked as an analytical chemist for the Queensland Government. His principal role was to assay minerals for the government. Working alongside F. M. Bailey, the keeper of the museum's herbarium, their shared expertise skewed the collections. Through the museum, they provided expert assistance by scientific staff and access to library volumes and displays for identifying ores, pests and plants. This aligned the museum's aims closely with the government's, advancing Queensland's status and economic viability. From 1888, the museum stored the state's standard weights and measures (Mather 1986, 226). These were not just conveniently housed but co-located with specimens, adding to the perceived utility of the museum.

Limited series of cultural materials were displayed, confusing the different communities' artefacts through grouped displays of 'weapons' or 'tools'. The Torres Strait Islands, formally incorporated into the state in 1879, were identified differently, as were the islands of the Pacific. In 1884, when Britain established a protectorate over the Southern and Eastern parts of New Guinea, naming it British New Guinea, missionaries and government personnel engaged in its governance began to send collections to the museum.

The majority of items were quickly put on display in the style of the world's fairs displays. Whilst museum labelling was limited, the museum had a regular presence in newspapers and a column written by curator Charles de Vis, highlighting new acquisitions or items on display. Charles de Vis had been appointed Director in 1882; although trained as a churchman, he had a keen interest in natural history and an understanding of anthropology (having been Vice-President of the British Anthropology Society); under his direction, and despite limited funds, the cultural collections grew.

As the nineteenth century progressed, the museum developed in tandem with the state, subject to the political changes and the boom-and-bust economic cycle. These factors drove the acquisition of cultural collections (and the bias towards the Pacific). For example, the Pacific Island labour trade brought indentured Islanders to the region, a practice known as 'blackbirding', and the ships that travelled back and forth became a source for artefacts collected, confiscated or stolen en route.

Collecting cultural artefacts was opportunistic and seldom planned, and offers were rarely refused. On occasion, sums of money were expended by the Trustees to acquire collections, but more often than not, only limited funds were available. While most collections came in sporadically and could come from anywhere within the state or beyond, an exception to this was the steady stream of artefacts arriving from British New Guinea (1888–98). Sir William MacGregor, the first Administrator of British New Guinea, had negotiated with the Queensland colonial administration for the museum to store and display the 'Official Collection' for British New Guinea, as they had no expectation of developing a museum of their own. Within a decade, almost 11,000 items had arrived at the museum, permanently skewing the cultural collections away from First Nations collections to New Guinea and the wider Pacific (see Quinnell 2000; Torrence et al. 2022).

The sheer and unexpected quantity of material sent from New Guinea exacerbated the issues of limited gallery spaces and workrooms. The museum, purpose-built in 1879, was already too small by 1882, and visitors and specimens jostled for space on the gallery floors. In the final years of the William Street building, MacGregor's shipments overwhelmed the displays and adjacent rooms. As quantities of material were received, ethnology registers were drawn up to list items and labels added to objects; however, despite much effort by De Vis and other staff, the registers failed to keep up with the number of artefacts arriving. This set the stage for future problems in interpreting and caring for the cultural collections (see Davies 2022).

By the end of the century, the Queensland Government had passed *The Aboriginals Protection and Restriction of the Sale of Opium Act* of 1897. This instituted the roles of Protectors who had enormous control over the everyday lives of Aboriginal and Torres Strait Islander people in Queensland. The controls included the establishment of reserves and missions and the forced movement of people from their ancestral Country and out of towns. As part of their duties, the Protectors travelled widely, and W. E. Roth, appointed to the role in the North, and Archibald Meston, based in the South, collected objects as they travelled (see Robins 1986; Price et al. 2021). Roth made a large collection of objects which he eventually sold to the Australian Museum which also published his research, this caused a furore as many believed the collection should have remained in the State. Meston regularly transferred small selections of objects to the museum and, as a former journalist, was more likely to write up his activities as commentary in the local newspapers. Their roles as Protectors gave them high levels of control over Aboriginal and Torres Strait Islander lives and they exerted this control in different ways, including removing objects and ancestral remains from the country (see Turnbull 2015 for a summary of their collection in relation to human remains).

In this first establishment phase, Queensland Museum's collection demonstrates partiality towards natural science collections. The acquisition of cultural material was heavily influenced by Queensland's participation in world's fairs and was made in response to offers of objects rather than having a strategic approach to building a 'representative' collection.

A new building – review and revision

Although rich in natural resources, Queensland remained a state of pastoralists. Until the late nineteenth century, its economy was vulnerable to drought (1866, 1890, 1898–1903), which affected wool production and sales and led to labour strikes (in 1891). The mining industry depended on infrastructure and only began flourishing in the 1890s when agriculture (sugar and wheat) also developed. The first museum was built during a period of economic growth, optimism and expansion for the new colony. Subsequently, renewed economic constraints led to a contraction in museum staff, reduced funding and prevented launching an architectural competition for a new museum.

Just before the turn of the century, in 1899, the museum moved again and took over the old Exhibition Building (erected as an exhibition and concert hall in 1891) in Bowen Hills. Now, on the fringes of the city, visitation decreased, and the next 50 years again saw cycles of intense investment and then negligence. In 1901 the states of Australia were united under the Federation which transformed Queensland from colony to State, and its participation in world's fairs ebbed (McKay 1997). As the century began the museum was once again in dire straits, De Vis was ageing and the staff was reduced. On De Vis' retirement a temporary director was appointed and the museum transitioned into a phase of severe decline. The newspapers lamented the state of the museum and, in response, Premier William Kidston initiated an external review in 1910. This was probably sped up due to the imminent return of major donor Sir William MacGregor to the region, this time to serve as State Governor.

The inspection was carried out by Robert Etheridge, then Curator (Director) of the Australian Museum in Sydney, and his report was detailed and scathing (Etheridge 1910). Trained as a palaeontologist and with a keen interest in ethnology, having set up a department in the Australian Museum (Walsh 1981), he brought a wealth of experience of museum practice with a broad disciplinary perspective. His report provides the first professional review of the museum and its functions and a detailed critique of the displays of cultural material.

In his report, Etheridge depicted an institution that was chaotic and mismanaged, with poorly interpreted displays that were overwhelmed with specimens. The skeleton staff lacked museological expertise, direction and subject knowledge. His report criticised the lack of labels, unscientific organisation and the inclusion of objects that, Etheridge argued, were not 'applicable' to a museum (Etheridge 1910, 5). On the ground floor, the Queensland Hall included displays of the Aboriginal collections. Located just inside the public entrance, he noted that these were presented in the 'International exhibition style', which 'however pleasing to the eye ... is totally unscientific' and 'forbids proper labelling' (Etheridge 1910, 4).

At the time of the move to Bowen Hills, the New Guinea collections were afforded a large gallery area for display, at the eastern end of the first floor. Co-located with other Pacific and overseas collections, only a portion could be accommodated, and many artefacts remained behind the scenes. Cases contained dense displays and objects adorned walls or sat in the rafters, making them inaccessible to staff or for more detailed study. This collection, due to its national and international significance, became a focus of the report and Etheridge's ire:

> New Guinea. Occupying by far the greater part of the Eastern side of the Gallery floor (about fifteen or more cases) and wall space is the fine MacGregor Collection. The cases are crammed to repletion [sic] the specimens roughly sorted, and not a label! Of what possible use is such a display? Just one half the number properly arranged and labeled [sic] would reveal to the onlooker the arts and manufactures of the British New Guinea natives in a manner not hitherto attempted. Along the walls are 'trophies' of speads [sic], arrows, etc. all unprotected as in the case of similar Australian objects ... An integral part of this collection is a magnificent series of Tapa cloths placed on the bare western wall of the Gallery, with one general label 'British New Guinea'. This should not be. I know of no other samples from New Guinea equal to these, and in all probability such a series will never be obtained again – they are worthy of better care. (Etheridge 1910, 6)

Although condemnatory, Etheridge's report was also fair and recognised how little the skeleton staff could accomplish when they had to double as cleaners and attendants (Mather 1986, 47–8). Premier Kidston responded by appointing Dr Ronald Hamlyn-Harris as Director, who reported to Kidston via the Colonial Secretary (the Board of Trustees had been disbanded in 1907). The registration processes were overhauled, and a new research publication series, *Memoirs of the Queensland Museum*, replaced the *Annals* that had petered out in the 1890s. Funds were made available to refit the galleries and redisplay cases.

Hamlyn-Harris quickly made an impact and completed a redisplay of the New Guinea collection which gained praise from MacGregor (Knowles and Curtis 2022, 450). He updated the taxidermy displays and installed the 'Aboriginal campsite diorama', which opened in January 1914. Dioramas were being used by other museums, and the Queensland Museum was a late adopter of the display technique (see Russell 2001 for a historiography and critique of the use of these display methods in Australian museums). Although racist, historicising and homogenising Aboriginal culture, it became a popular feature of the museum and remained on display until 1985, when the museum was closed before its relocation (Rowlands 2011, 205–6; Burden 2019, 746–8).

Hamlyn-Harris proactively sought Aboriginal and Torres Strait Islander artefacts, and as early as 1911 sent a circular to police and government officials across the state outlining the urgent need to develop the collections. He explained, 'Since the Aboriginal Tribes are fast dying out, every effort should be made to acquire those symbols of the life of the original Australian inhabitants' and that items should be sent to the museum before 'being entirely lost to us' (Quinnell 1986, 208–9). Although several correspondents replied that it was already too late, as traditional implements were no longer being made, it led to a second wave of collections coming in from the police, government officials and residents who were based in more remote areas, which Burden notes 'saw the collection [of Aboriginal artefacts] nearly double during his tenure' (Burden 2014, 87).

Hamlyn-Harris's tenure was short-lived and the government's financial support did not endure. After his resignation in 1917, his successor, Heber Longman, continued the programme of redisplay and maintained typological groupings with general labels, but expanded displays to include contextual photographs (Knowles and Curtis 2022). Longman focused on re-registration and the implementation of the new registration systems and records for the cultural collections. He accepted what was offered and encouraged the expansion of the physical anthropology collection (Quinnell 1986, 211–12). Gradually, even 'passive collecting' ground to a halt, and the ongoing maintenance of collections led to exchanges to other museums, and also the refusal to acquire important Queensland collections (Quinnell 1986, 212–14).

In 1933, the museum once again came under external review. The British Museums Association, funded by the Carnegie Trust, reviewed museums across the empire and published a volume on Australia and New Zealand (Markham and Richards 1933). The report compared civic population sizes, collections' scope and scale, staffing and income across the Empire, through a series of volumes. The Queensland Museum, when compared in a table to museums in Adelaide, Bristol, Hull, Bradford, Johannesburg, Cape Town and Cardiff, was singled out for special commentary:

> This table clearly highlights the poor finances of Brisbane, which apparently has a smaller museum and art gallery income than any similarly sized town in the Empire, even though it has state resources to draw upon ... The income problem alone may account for the comparative poorness of the museums at Brisbane and Perth. (Markham and Richards 1933, 10)

The report highlighted the continued lack of Trustees (and therefore lack of direction, supervision and reporting) and recorded that the staff was too small for the size of collections. However, the paramount concern was the museum itself: 'Possibly the most unsuitable museum building in the Commonwealth is that at Brisbane, which is a positive fire-trap' (Markham and Richards 1933, 27).

From curios to culture – the postwar period and the establishment of anthropology

In 1955, Museum Director George Mack organised a special exhibition on the Centenary of the Queensland Museum, which traced the history of the museum and featured directors, curators and collectors (Mather 1986, 82). The following year, Mack wrote a short history of the museum and highlighted the ongoing neglect that dogged the New Guinea collection, both front of house and behind the scenes (Mack 1956). He argued that a dedicated curator would better serve the collection (Mack 1956, 118). The issues Mack identified were enumerated against the backdrop of the campaign for and establishment of an Anthropology Museum at the University of Queensland. The museum's founding collection was the work of L. P. Winterbotham and had been assembled by him over a period of 20 years. Initially focusing on Queensland, he expanded his network to include the Pacific. Winterbotham's collecting flourished as Queensland

Museum's diminished. In effect, the Anthropology Museum demonstrated that many felt the need to fill a void created by the Queensland Museum's lack of staff, space and subject matter expertise (see Leo 2008 for a history of the collection and early years of the museum). Even with the appointment for the first time, in 1966, of a Curator of Ethnography, Eleanor Crosby – who immediately changed the title to Anthropology – the role was necessarily focused on improving the care of the collections within the museum rather than seeking out new acquisitions.

During this post-war period, global politics set a new context for cultural collections in museums. The UN Declaration on the Granting of Independence to Colonial Countries and People (1960) was closely followed in the Pacific by New Zealand's release of Western Samoa (1962). In 1970, Tonga and Fiji became the first British colonies in the Pacific to secure independence and in 1975, Australian-administered Papua New Guinea became a nation-state. Political change drove a review of the role of museums in relation to their local and overseas communities. In Papua New Guinea, for example, during the years and negotiations leading up to independence and the transfer of power, the first nationals were appointed to the museum's Board of Trustees, and there was a gradual transfer from expatriate to local staff.

After little more than a year, Crosby resigned from Queensland Museum and was succeeded by Michael Quinnell, who continued the identification and co-location of cultural material, with a particular focus on the New Guinea collection. In 1967, a referendum was passed to enshrine constitutional change to recognise Aboriginal and Torres Strait Islander peoples for the first time in the census; 90.77 per cent of the population voted in favour of the change, the most successful referendum in Australian history (AIATSIS 2021).

The emerging postcolonial world influenced global debates on museum collections, custodianship and Indigenous rights. In Australia, the Commonwealth Government commissioned the Pigott Report on museums in Australia (Pigott 1975). A turning point for museums in considering the future was the UNESCO regional seminar on the 'Role of Museums in Preserving Indigenous Culture', held in Adelaide in 1978. One of many responses to the call for change included the establishment in 1979, of the Conference of Museum Anthropologists (COMA), a forum for sharing and advancing more inclusive practices (see Stanton 2011 and Allen 2021 for detailed reviews of this period). By 1980, UNESCO's intergovernmental committee for promoting the return of cultural property to its countries of origin or its restitution in case of illicit appropriation had released a report (UNESCO 1980). Former colonies established new national museums using colonial collections and presented them as spaces of unity speaking to emerging national identities. In 1975, to commemorate Papua New Guinea's transition to Independence, the Australian Government funded the building of the National Museum and Art Gallery to house the collections of former Colonial Governors and Officers.

In 1970 Queensland Museum stopped acquiring ancestral remains. Aboriginal and Torres Strait islander visitors and staff began to draw attention to the problems with having secret or sacred artefacts and ancestral remains on display, and some items were removed to storage areas. The first discussions on the return of cultural heritage focused on the New Guinea collections. At the 1970 Conference of Australian Museum Directors, a request was made for the return of the Official Collection to Papua New Guinea, and the attendees provided the recommendation 'that a fully representative selection of the MacGregor Collection be returned to the Papua New Guinea Museum' (Craig 1996, 203). This prompted the Queensland Museum to ask the Queensland Government to seek legal opinion; the Solicitor-General concluded that Papua New Guinea would have a claim. In 1974, Queensland Premier Joh Bjelke-Petersen announced that the collection would be returned (Quinnell 2000, 96). This announcement, although largely ignored in the Queensland press, was reported in the *Papua New Guinea Post-Courier* (1974) under the headline 'Artefacts will be returned'.

Throughout the 1970s, Quinnell began a series of changes to the displays to develop a modern display of Melanesian art. As older cramped displays were dismantled, the quantity of unregistered material stored on display came to light, and documentation had to be further revised. These changes had a limited effect on how the museum was viewed by most visitors; the large wooden cases limited display options and displays remained in similar groupings to those installed in 1899. Behind the scenes, the storage and staffing of the cultural collections (Archaeology and Anthropology) were changing radically. The departure of the art gallery in 1977 eased space in the storage, and the collections could be co-located and sub-divided, making documentation easier, although access remained difficult.

For the next decade or more Quinnell focused on facilitating access and assessment to the collection by the staff of the Papua New Guinea National Museum and Art Gallery. A decision was made to split the collection, with the majority of the collection to be returned (60 per cent) and the remainder gifted to the Queensland Museum.

Over the next 30 years, work to photograph, document and then select and transfer objects was ongoing. The museum recruited more staff to manage the Aboriginal and archaeological collections, including Richard Robbins who led the Australian Anthropology and Archaeology Section, supported by Norma Richardson and Judith Wassell (nee Bartlett). They increased capacity and brought new connections to Aboriginal community organisations such as FAIRA (Foundation for Aboriginal and Torres Strait Islander Research Action) paving the way for the recruitment of First Nations staff (Quinnell 1986, 216–17).

By the 1980s, when the first items had been transferred to Papua New Guinea National Museum and put on display, the Queensland Museum was looking towards a new home in the recently built cultural precinct on the south bank of the Brisbane River. The precinct was at the west end of a large-scale redevelopment of a former industrial area which had been cleared to host the World Expo '88. Once again, the Queensland Museum's entanglement with world's fairs and exhibitions reasserted itself (Bennett 1991). The move presented an opportunity to rethink the displays and interpretation. In preparation for the move, Quinnell curated a suite of cases, *Melanesia: People and traditions of the south-west Pacific*, and the Aboriginal display *Jirrbal: A rainforest people*, curated by consultants from James Cook University in collaboration with the Jumbun Community supported by First Nations preparator Carolyn Cowell. These exhibitions had an expected lifespan of five years (Robins 1986, 116).

A new museum and a new era

A comprehensive history of the Queensland Museum was written on the eve of the museum's wholesale removal from the old Exhibition Building in Bowen Park to the new cultural precinct on South Bank in 1986 – a celebration of 125 years of collecting, exhibitions and research (Mather 1986). The collection of essays detailed and celebrated the achievement of staff in overcoming problematic buildings, inconsistent financial support and limited staffing. However, the inward focus ignored the museum's wider social context.

The new building co-located stores and staff offices, and the cultural collections were placed together accessible in one storage room, with an associated 'Anthropology lab', which stored related documentation and collection photographs and provided a space for staff and visitors to view collection items. Two First Nations cadets, Lori Richardson and Shane

Rawson, were employed, followed by Tina Baum, the first Aboriginal Curator, who was succeeded by Michael Aird in 1995. This marked a new era; while previous curators had welcomed visitors with cultural and familial ties to the collections, the broader social networks that First Nations staff provided encouraged more people generally to visit back of house. Before Baum's appointment, less than 25 Aboriginal people would visit the laboratories or stores in the museum each year. Baum increased this figure to 800 people per year, and Aird increased it further, consistently maintaining 1,200 or more visits annually.

The new museum offered different opportunities. Despite the lack of permanent display there emerged innovative responses which were sector-leading and locally impactful. The performance and associated exhibition *You Came to My Country and You Didn't Turn Black* (20 August to 8 September 1990) marked a significant change of direction. The performance was originally conceived and directed by Sue Rider and was created from the poems of Oodgeroo Noonuccal to be performed at the Adelaide Festival in 1984. Six years later, through the support of Curator Judith Wassell, the performance was further developed by including poems by Maureen Watson and an associated exhibition of contemporary artworks by Queensland First Nations artists. The play 'explored one woman's vision of her people's joy and sorrow as they strived to come to terms with alienation with a land which was once theirs' (Wassell 1993, 42), and the exhibition comprised 145 artworks which responded to these themes. The exhibition included works for sale by Richard and Marshall Bell, Gordon Bennett, Fiona Foley, Ron Hurley, Gordon Hookey and Judy Watson. The performance coincided with the Conference of Museum Anthropologists (COMA) meeting in Brisbane under the theme *Facing Responsibility* (Wassell 1993, 2).

The significance of this event and its impact on participants and the museum was acknowledged in 2015 when Michael Aird and Mandana Mapar staged a photographic exhibition for the Queensland Museum. The exhibition *This is My Heritage*, took its title from a poem included in the play. The exhibition curators reflected that 'The political nature of the poems and the responses from the visual artists were integral to the overall success of the project. At the time, it was a major coup and a notable change in direction for the Queensland Museum. The project captured the mood of the era …' (Aird and Mapar 2015, 7).

The 1990s was a period of sweeping change. With new spaces to work in, hold forums and gatherings, the lack of permanent galleries was creatively tackled with an increasing number of educational programming, forums, outreach and special exhibitions. Michael Aird, who had recently been employed to curate the temporary exhibition based on studio portraits of Aboriginal people, *Portraits of Our Elders* (1991), was employed as Curator from 1995–2000. The success of the exhibition led to a publication and touring exhibition (Aird 1993). Aird brought an understanding of the relationships that were captured in the portraits and highlighted the agency of Aboriginal people in commissioning photographs of themselves (see Besley 2015). Having demonstrated the significance of the photographic collection to First Nations audiences, Aird scoured the stores and the wider museum to bring together any image of an Aboriginal person or Torres Strait Islander in any medium, whether exhibition prints, photocopies, negatives or albums, to augment the photographic or EH (ethnohistorical) collection. He then compiled a series of ring-binders where a copy of every image was placed in a plastic slip alongside its collections record and made available by region to any visitor to the Anthropology lab.

Aird had a specific audience to serve: the Aboriginal and Torres Strait Islander communities. Through providing folders of photographs and related information, Aboriginal visitors could review the entire collection, identify ancestors, family or friends and then take photocopies home with them. The folders were taken out of the museum for special events, particularly NAIDOC day celebrations (formerly an acronym but now the name of a day when Aboriginal and Torres Strait Islander culture is celebrated nationally) at nearby Musgrave Park, where a tent and tea urn were set up, and people could thumb through the folders and share images and stories over a cup of tea.

Although the focus of work on the Pacific collections remained on the return of the Official Collection to Papua New Guinea, the museum began to consider the wider Pacific collections differently with the employment of Imelda Miller, an Australian South Sea Islander. Over several years, Miller compiled a guide to the Kastom collection, revisiting the Pacific collections associated with blackbirding and refocusing it to align with Australian South Sea Islander history. Through her work, the collections shifted from a colonial museum lens, and instead, the significance of their association with descendant communities was prioritised. This re-attribution reasserted ownership and an ancestral kinship between the newly recognised Australian South Sea Islander community (the Commonwealth Government recognised Australian South Sea Islanders in 1993 followed by the Queensland Government in 2000). As the Kastom collection alternative narratives and truths emerged, Miller's ongoing commitment to this collection and her community has led to a series of exhibitions and activities which centre on Australian

South Sea Islanders' history. These have been hosted across the cultural precinct weaving new perspectives into the fabric of old institutions.

As new narratives were prioritised through reviews of collections documentation practices, exhibitions and events, the repatriation of ancestral remains also began in earnest. The museum did not wait for the legislation that asserted Aboriginal and Torres Strait Islander rights to their ancestors which came a decade later (Queensland Government 2003). Instead Aird, supported by the museum, facilitated a series of returns. He was keen to see ancestral remains return home but demonstrated an understanding of the complexity and emotions involved. For example in 1996, when he arranged the return of a child's skull to the Napranum Community (Cape York) he also agreed to the request from a senior woman, Alice Mark-Andrews, to wear a shell necklace from the museum at the burial ceremony. In recounting this story, Aird describes that although the necklace was returned only temporarily, this was an act of 'cultural knowledge being returned to a community' (Aird 2002, 306).

Aird brought this approach to collecting new artefacts as well; his small-scale exhibitions often focused on contemporary works and the politics of the time. *Wearing Culture: Images and Objects from an Artists' Meeting* (1998) followed two years after *Presenting Culture*, a public forum hosted by Queensland Museum. This responsive interaction between museum and community supported dialogue and critique. Aird sought to acquire new collections that would be meaningful to Aboriginal and Torres Strait Islander people and engender a different type of engagement with the museum. He regularly visited Queensland Aboriginal Creations (QAC), the government-funded arts shop located across the Brisbane River. At this time, training for Aboriginal artists was only beginning; Aird's foresight in documenting this social history means that the Queensland Museum has a unique collection of early contemporary art from significant Aboriginal and Torres Strait Islander communities as well as QAC shop signs and associated ephemera. As Stolte notes Aird 'saw each object as a representation of someone's personal story of Indigenous expression' (Stolte 2020, 5). In acquiring these artworks Aird began to transform the shape of the collection from a frequently extractive Western perspective to an Aboriginal one.

Following Aird's tenure, Olivia Robinson was appointed Senior Curator of Aboriginal Studies and was followed by Trish Barnard as Senior Curator of Indigenous Studies. They both continued the temporary exhibition programme and associated activities but consciously asserted a female perspective. This, they argued, went some way to address the significant silencing of women through the inherent biases in the collections that were mostly collected by men and focused on male objects. Through their critical appraisal of the collections and the exhibitions and activities they undertook, they fostered 'a feminine environment [that] has empowered Indigenous women to express their cultural identity and gender perspective through art and exhibitions, which has led to a reaffirmation of ownership of cultural heritage' (Robinson and Barnard 2007, 35). The exhibition *The Birthing Circle*, developed with Yugambeh Museum, Language and Heritage Research Centre and shown at the Queensland Museum (2005–7), traced four generations of Yugambeh women and how their lives were influenced under colonial governance 'taking a spiritual experience [birthing] and transforming it into a clinical procedure' (Robinson and Barnard 2007, 42).

Robinson was the lead curator for *Dandiiri Maiwar*, the first dedicated First Nations gallery which opened in December 2004. Described as an Aboriginal and Torres Strait Islander Cultures Centre, it focused on three themes: Land, Language and Lore. The gallery, developed by Indigenous architect Kevin O'Brien, comprised three different spaces: a central exhibition area of six pods which highlighted topics or a cultural or regional group; an adjacent area, the open access collection, which was filled with displays of object types showing regional variation and provided a window on the wider collections; the third intended to be a staffed area with a large yarning table, where folders of photographs and collections databases could be accessed. Sadly the centre was never permanently staffed and so, despite major investment, its potential was not realised. An institutional understanding of the interplay of the three galleries was gradually lost as staff changed and the open access collection was removed in 2012. At around the same time Aird, a strong advocate for this approach to collections access, was re-employed to develop the concept for a revised Aboriginal and Torres Strait Islander gallery.

While permanent galleries for Pacific and First Nations cultural material have remained elusive, small and large-scale exhibitions have figured from time to time. Among these, one case displays have marked the 40th anniversary of independence in Papua New Guinea (*Wan Kantri*) and highlighted popular tableware of the late twentieth century that featured Aboriginal motifs and designs. From 2017–18, Alethea Beetson led a collaboration with Digi Youth Arts, an award-winning long-term residency for young people at the museum, resulting in two exhibitions, one of which *660: Calling Home*, was a response to the ancestral remains still held by the museum. Other creative works included a music album and performance, three short films and a play *Anthrop-apology*. A recent exhibition *Island Futures: What lies ahead for Zenadth Kes?* (2021) was installed on the entry level, where the new permanent gallery had been proposed. This exhibition provided an opportunity to highlight issues that face Torres Strait Islander communities today, including climate change, pollution and cultural change.

At a national level, guidance was changing again. The Australian Museums and Galleries Association (AMAGA) had commissioned and consulted on its *Indigenous Roadmap*, which was completed in 2018 and provided pathways for enhancing Indigenous engagement in the sector (Janke 2018). By this time, many museums were adopting Reconciliation Action Plans, a programme requiring organisations to build positive relationships between Indigenous and non-Indigenous people.[1] Queensland Museum launched its first *Reflect* RAP in November 2019 (Queensland Museum 2023, 15). The latest *Innovate* RAP was launched in 2023 and acknowledges the Queensland Government's *Path to Treaty Bill 2023*.

Conclusion

In October 2023 Queensland roundly rejected the national referendum on the Indigenous voice to parliament, polling the lowest percentage in favour (less than 32 per cent) of all the states across Australia and the highest rejection of any electorate in the country, with 84 per cent of the electoral division of Maranoa in the Southern outback voting 'No' (Biddle et al. 2023, 14). The referendum came out of the *Uluru Statement from the Heart* issued in 2017 at the close of a constitutional convention of 250 Aboriginal and Torres Strait Islander delegates. The statement has since been summarised as a series of actions and reforms under the slogan, 'Voice, Treaty, Truth'. Voice is a call for constitutional change to establish an Indigenous voice to parliament (which the referendum sought to achieve); Treaty, the negotiation of treaties between First Nations people and the government; and Truth, a truth-telling of history. The Uluru statement summed up the journey as follows: 'In 1967 we were counted, in 2017 we seek to be heard ... We invite you to walk with us in a movement of the Australian people for a better future' (Referendum Council 2017).

In the early days, Queensland Museum was part of building a national (later state) identity and a source of civic pride. The museum's close association with Queensland's desire to market itself on the world stage through world's fairs was one tool for taking Queensland's place in the Empire and promoting the State as a place of opportunity for potential migrants. In the early years, it provided a central location for the education and entertainment of households in a relatively small town in a remote colony. At its heart, the museum was built for the new settlers, based on principles of extraction of knowledge and resources from its Indigenous inhabitants. For First Nations, it was a place that gained its collections through the violent dispossession of people, their culture and country.

Politics, economics and legislation have continued to shape the museum's activities and continue to shape Queensland's relationship with its First Nations and near Pacific neighbours. This context still impacts the collection, research, display and interpretation of cultural material in the Queensland Museum, but a cultural context, a vocal First Nations voice, now questions the purpose, right and authority of the museum with respect to its collections. Where is the future of ethnography in this institution in the current social and political context?

Today, the Queensland Museum has a First Nations Director who leads curatorial direction and access. Dr Bianca Beetson was appointed in 2023 and, for the first time, this role places a First Nations person on the Executive team. The repatriation unit has a dedicated staff and received AU$4.5 million in dedicated funding from the Queensland Government in 2023 (Guenzier 2023).

The heady days of a tea urn in Musgrave Park, in which community members, staff and the Queensland Museum Aboriginal and Torres Strait Islander Consultative Committee (QMATSICC, established in 1997) hung out with the photographic collections among the NAIDOC day crowds seem like another era, when Indigenous and non-Indigenous staff readily worked together to innovate change. In many ways it was a different time; it relied on personal commitments and connections rather than strategic institutional support, which ultimately led to its demise. However, its legacy has been key to current success, a better-informed community that understands what the museum holds and wishes to hold the museum and its guardianship to account. And it is a community that wants to be part of decision-making and have story-telling within its walls. However, it remains frustratingly uneven in a lack of strategic, sustained support for First Nations staff and open access for an informed First Nations community.

As I stood and listened to the speeches in *Dandiiri Maiwar*, as community members came to mark its closure, I doubted that the commitment voiced – that a new gallery would open within two years – could be met. The political context meant most arts and culture funding was focused on commemorating World War I, and natural history projects continued to be the main attractor for private companies, particularly those who wished to improve their environmental credentials. Engaging with Indigenous cultures was politically charged and concerns about failure or critique kept substantial funders at arm's length. In the end my concerns were well founded: funding was not secured and the two-year commitment passed. However, the 10 years that have elapsed since have not been static. Instead, the museum's growing support of large-scale temporary exhibitions prominently located on the entry-level has led to new outcomes; rather than being static long-term displays they are responsive and provide deep engagement with specific regions or communities. Finally, the failure of the referendum on constitutional reform has been tempered by the Queensland Government's commitment to Treaty and truth-telling (Queensland Government 2003).

For Queensland Museum, it is perhaps the cultural precinct that serves it best. The unique co-location with other institutions, which provides access to First Nations culture across all art forms and showcases the past, present and future, tempers the traditional audience expectations of an encyclopaedic museum which disassociates nature and culture. As long as the museum continues to grow its First Nations workforce and address its role in historic harm it could become a safer space for Aboriginal and Torres Strait Islander staff and visitors, with renewed opportunities for access greater than the levels seen in the 1990s. It must continue to prioritise access to collections and innovate in its curation, interpretation, exhibition and return of the artefacts it holds on behalf of First Nations, Australian South Sea Islander and Pacific Communities.

Note

1 See https://www.reconciliation.org.au/the-rap-framework/ (Accessed 30 August 2024).

References

AIATSIS. 2021. 'The 1967 Referendum'. Australian Institute of Aboriginal and Torres Strait Islander Studies. https://aiatsis.gov.au/explore/1967-referendum.

Aird, M. 1993. *Portraits of Our Elders*. South Brisbane: Queensland Museum.

Aird, M. 2002. 'Developments in the repatriation of human remains and other cultural items in Queensland, Australia'. In *The Dead and Their Possessions: Repatriation in principle, policy and practice*, edited by Cressida Fforde, Jane Hubert and Paul Turnbull, 303–11. London: Routledge.

Aird, M. and M. Mapar. 2015. *This is My Heritage*. Brisbane: Queensland Museum.

Allen, L. 2021. 'Museum collections and their legacies'. In *The Oxford Handbook of the Archaeology of Indigenous Australia and New Guinea*, edited by Ian J. McNiven and Bruno David, 101–212. Oxford: Oxford Academic. https://doi.org/10.1093/oxfordhb/9780190095611.013.7.

Bather, F. A. 1894. *Some Colonial Museums*. London: Museums Association.
Bennett, T. 1991. 'The shaping of things to come: Expo '88'. *Cultural Studies* 5 (1): 30–51. https://doi.org/10.1080/09502389100490031.
Bennett, T. 2005. 'Civic laboratories: Museums, cultural objecthood and the governance of the social'. *Cultural Studies* 19 (5): 521–47. https://doi.org/10.1080/09502380500365416.
Besley, J. 2015. '"Looking past": The work of Michael Aird'. *reCollections* 10 (1). https://recollections.nma.gov.au/issues/volume_10_number_1/papers/looking_past.
Biddle, N., M. Gray, I. McAllister and M. Qvotrup. 2023. *Detailed Analysis of the 2023 Voice to Parliament Referendum and Related Social and Political Attitudes.* Canberra: The ANU Centre for Social Research and Methods. Accessed 31 August 2024. https://csrm.cass.anu.edu.au/research/publications/detailed-analysis-2023-voice-parliament-referendum-and-related-social-and.
Burden, G. 2014. '"A centre of scientific work": Ronald Hamlyn-Harris and ethnography at the Queensland Museum 1910–17'. *Journal of Australian Studies* 38 (1): 87–102. https://doi.org/10.1080/14443058.2013.865660.
Burden, G. 2019. '"The adherent taints of civilisation": Entangled artefacts and the Queensland Museum, c.1880–1920'. *History Australia* 16 (4): 733–49. https://doi.org/10.1080/14490854.2019.1670073.
Craig, B. 1996. 'Samting Bilong Tumbuna: The collection, documentation and preservation of the material cultural heritage of Papua New Guinea'. PhD dissertation, Flinders University of South Australia.
Daintree, R. .1873. *Queensland, Australia: Its territory, climate and products, agricultural, pastoral and mineral, etc etc, with emigration regulations, illustrated with maps and plates.* London: G. Street.
Davies, S. M. 2022. 'Becoming a collection (1892–2001): Challenges and practices in the curation of the Official Collection at the Queensland Museum'. *Memoirs of the Queensland Museum – Culture* 13: 27–74.
Etheridge, R. 1910. *Report on the Queensland Museum.* (QMPP61). Brisbane: Queensland Museum Library Archives.
Guenzier, J. 2023. '"It's quite disturbing": Uncle Bob Weatherall calls on Queensland Museum to provide register of stolen cultural material'. *National Indigenous Times*, 29 September. Accessed 31 August 2024. https://nit.com.au/29-09-2023/7851/uncle-bob-weatherall-calls-on-queensland-museums-to-provide-register-of-stolen-cultural-material.
Janke, T. 2018. *First Peoples: A roadmap for enhancing Indigenous engagement in museums and galleries.* Canberra: Australian Museums and Galleries Association.
Knowles, C. and N. Curtis. 2022. 'Exhibiting the MacGregor Collection'. *Memoirs of the Queensland Museum – Culture* 13: 443–68.
Leo, D. 2008. 'An ark of Aboriginal relics: the collecting practices of Dr LP Winterbotham'. In *The Makers and Making of Indigenous Australian Collections*, edited by Nicolas Peterson, Lindy Allen and Louise Hamby, 29–60. Melbourne: Melbourne University Press.
Mack, G. 1956. 'The Queensland Museum 1855–1955'. *Memoirs of the Queensland Museum* 13: 107–24.
Markham, S. F. and H. C. Richards. 1933. *A Report on the Museums and Art Galleries of Australia to the Carnegie Corporation of New York*. London: The Museums Association.
Mather, P. 1986. *A Time for a Museum: The history of the Queensland Museum 1862–1986*. Brisbane: Queensland Museum.
McKay, J. 1997. *'A Good Show': Colonial Queensland at international exhibitions.* Brisbane: Queensland Museum.
McKay, J. 2004. *Showing Off: Queensland at world expositions 1862–1988*. Rockhampton: Central Queensland University Press.
Museums Australia Inc. 1993. *Previous Possessions, New Obligations: Policies for museums in Australia and Aboriginal and Torres Strait Islander Peoples*. Canberra: Museums Australia Inc.
Museums Australia Inc. 2005. *Continuous Cultures, Ongoing Responsibilities*. Canberra: Museums Australia Inc.
Papua New Guinea Post-Courier. 1974. 'PNG artifacts will return to us as gifts'. *Papua New Guinea Post-Courier*, 4 October: 15.
Pigott, P. H. 1975. *Museums in Australia, 1975. Report of the Committee of Inquiry on Museums and National Collections including the Report of the Planning Committee on the Gallery of Aboriginal Australia.* Canberra: Australian Government.

Price, S., L. Allen and C. Knowles. 2021. '"The (not-so)sacred Ibis": Archibald Meston, the colonial collector, and the Queensland Museum'. *Memoirs of the Queensland Museum – Culture* 12: 73–121. https://doi.org/10.17082/j.2205-3239.12.1.2021.2021-04.

Queensland Government. 2003. Aboriginal Cultural Heritage Act 2003. https://www.legislation.qld.gov.au/view/pdf/inforce/2016-09-27/act-2003-079.

Queensland Museum. 2023. 'Reconciliation action plan'. Accessed 27 September 2024. https://www.museum.qld.gov.au/about/corporate-information/reconciliation-action-plan.

Quinnell, M. 1986. 'Peoples and lifestyles: Anthropology'. *Memoirs of the Queensland Museum* 24: 199–220.

Quinnell, M. 2000. '"Before it has become too late": The making and repatriation of Sir William MacGregor's Official Collection from British New Guinea'. In *Hunting the Gatherers*, Volume 6, edited by M. O'Hanlon and R. L. Welsch, 81–102. Oxford: Berghahn.

Referendum Council. 2017. *Final Report of the Referendum Council*. Accessed 31 August 2024. https://ulurustatemdev.wpengine.com/wp-content/uploads/2022/01/Referendum_Council_Final_Report.pdf.

Robins, R. 1986. 'Queensland Museum'. *Australian Archaeology* 23: 115–16.

Robinson, O. and T. Barnard. 2007. '"Thanks, but we'll take it from here": Australian Aboriginal and Torres Strait Islander women influencing the collection of tangible and intangible heritage'. *Museum International* 59 (4): 34–45. https://doi.org/10.1111/j.1468-0033.2007.00621.x.

Rowlands, S. 2011. 'Entangled frontiers: Collection, display and the Queensland Museum, 1878–1914'. *Journal of Australian Colonial History* 13: 183–206.

Russell, L. 2001. *Savage Imaginings: Historical and contemporary constructions of Australian Aboriginalities*. Kew: Australian Scholarly Publishing.

Stanton, J. E. 2011. 'Ethnographic museums and collections: from the past into the future'. In *Understanding Museums: Australian museums and museology*, edited by D. Griffin and L. Paoissien. Canberra: National Museum of Australia. Accessed 31 August 2024. https://tinyurl.com/38we7nhy.

Stolte, G. 2020. *Agency and Legacy: A critical essay*. Brisbane: University of Queensland. Accessed 27 September 2024. https://anthropologymuseum.uq.edu.au/files/5231/QAC%20Agency%20and%20Legacy.pdf.

Torrence, R., E. Bonshek, A. Clarke, S. M. Davies, J. Philp, M. Quinnell and C. Knowles. 2022. 'Collecting for the nation: Sir William MacGregor and the Official Collection'. *Memoirs of the Queensland Museum – Culture* 13: 1–25.

Turnbull, P. 2015. 'Australian Museums, Aboriginal skeletal remains, and the imagining of human evolutionary history, c. 1860–1914'. *Museums and Society* 13 (1): 72–87.

UNESCO. 1980. 'Report of the Intergovernmental Committee for Promoting the Return of Cultural Property to its Countries of Origin or its Restitution in case of Illicit Appropriation. 1st session, Paris, 1980'. UNESCO, General Conference, 21st session, Belgrade, 1980. Accessed 27 September 2024. https://unesdoc.unesco.org/ark:/48223/pf0000041941?posInSet=1&queryId=49778cd7-3bce-40f2-b72e-6b5b15f03cc3.

Walsh, G. P. 1981. 'Etheridge, Robert (1846–1920)'. In *Australian Dictionary of Biography*. Accessed 1 September 2024. https://adb.anu.edu.au/biography/etheridge-robert-6117/text10489.

Wassell, J. 1993. 'You came to my country and you didn't turn Black'. *Bulletin of the Conference of Museum Anthropologists* 25: 42–8.

6

'We hate ethnography …': curating beyond description in a post-colonial museum

Sean Mallon

The Martinique born novelist, poet and theorist Édouard Glissant (1928–2011), reflecting on the colonial experience in the Caribbean, memorably wrote, 'We hate ethnography … the distrust that we feel toward it is not caused by our displeasure at being looked at, but rather by our obscure resentment at not having our turn at seeing' (Glisson, cited in Britton 1999, 23). An analyst of his writings suggests he is referring to 'the inequality and lack of reciprocity of the relation, in which those who are seen cannot themselves "see" those who see them' (Britton 1999, 23). This reflection captures some of the tensions of the phenomena of the post-colonial museum, the involvement of some of its former 'subjects' in its development, and the ongoing debates around ethnographic displays and representation. As a curator and anthropologist of Sāmoan and Irish descent my entry into the Museum of New Zealand Te Papa Tongarewa (Te Papa) was part of this ongoing post-colonial project, and an initiative enabling indigenous Māori, Pacific and indeed all peoples represented in Te Papa to have their turn at 'seeing'.

The catalyst for this was the implementation in the museum's operations of an indigenous Māori concept called Mana Taonga – a concept that acknowledges the spiritual and cultural connections communities have to taonga and collection items.[1] In a practical sense, Mana Taonga accords rights to those with such connections, to participate in the care of their taonga or cultural treasures and to speak about and determine their display or other usage by Te Papa. It is a concept and a sharing of authority that applies not only to Māori but to any individuals, families and communities that have cultural property or are represented in the museum (McCarthy et al. 2013; Schorch and Hakiwai 2014; Schorch et al. 2016).

Questions about authority, and who represents whom, have been part of anthropological critique and debate since the 1960s. So much so that critics were compelled to comment that 'Anthropology has reached a stage at which even the critics of academic authority criticise each other's authority' (Nencel and Pels 1991, 2). The implementation of Mana Taonga as part of the Te Papa museum project aligned with developments in the anthropology of the 1980s and 1990s, where ethnography as a primary research method was '… disputed as a Western monopoly on ethnic narratives' (Clifford 1983; Clifford and Marcus 1986; Geertz 1988) and how from this debate emerged '… a reflexive and dialogic anthropology, aiming to more adequately relate these encounters' (Debary and Roustan 2017, 4–5). In the first decades of the twenty-first century, 'in anthropology and museums, the debate has now been displaced onto the issue of shared authorship, and even ownership, of ideas and collections' (Debary and Roustan 2017, 4–5).

As curators, any authority we have partly derives from our job titles, and the cultural cachet this title carries in the popular imagination because of its association with museums as sources of authoritative knowledge. But it does not precede our practice. We have to work at establishing authority and sharing it with our collaborators. My experience of working with Mana Taonga is that it is not a fixed idea or blueprint but, rather, a working concept, an enabler for authority to be shared and negotiated to suit particular circumstances and various types of museum activities, from exhibitions to publications and public programmes.

For the Te Papa project, establishing an institutional orientation to authority began with the decision to create and conceptualise a museum that includes a Pacific peoples' focus to start with. Before arriving at a point where curators conceptualise and present an idea for an exhibition, the establishment of authority took hold with the first Pacific community consultations, the hiring of interns to train for the new reimagined institution, and then the formation of the Pacific curatorial team.

In a classic anthropological sense, we the curatorial team members were 'participant observers' but our privilege, and the power dynamics among those we worked with, was unlike a stereotypical relationship between anthropologist or curator and the native peoples under study. When we as Pacific peoples were facilitating the exhibition-making process and curating our own people's stories and material culture, the issues and challenges of the participant observer took on a familiar but perhaps riskier form. There were different expectations around engagement because we were of the community (although the concept of community can be interrogated). We did not confer authority or scientific validity on our exhibitions merely by our presence as anthropologically informed curators. We were not there to observe and collect objective data nor were we 'bound to the retrospective chronicling of lives that are always on the brink of disappearing' (Ingold 2014, 393). We negotiated our work with all the participants that constituted the process. Our role was not to objectify, as anthropologist Tim Ingold argues, but '… to attend to persons and things, to learn from them, and to follow in precept and practice. Indeed, there can be no observation without participation – that is, without an intimate coupling, in perception and action, of observer and observed' (Ingold 2000, 108). There was no handbook for the work we were undertaking, just a guiding principle of Mana Taonga. Although, as time went on, we did develop a kind of virtual playbook, if I could call it that, a collective shared set of experiences that we could draw from of 'go to' strategies, case studies and principles that informed our future projects (Mallon 2016).

Exhibitions are a powerful expression of Mana Taonga because they activate community engagement in a sustained and highly visible way, bringing transparency and accountability to how the museum conducts its work. Exhibitions fill visitors' views of what makes up the DNA of the museum. They are the front-facing elements of the visitor experience, the frontier for onsite public engagement. An analysis of the museum's history of display has much to reveal about the staff, communities and politics at various points in time. It may chronicle the institution's changing philosophy, its public life and its growth or demise. Here, rather than focus on the formal qualities of the ethnographic displays we made, I look beyond processes of curatorial description towards processes of collaboration. I analyse the politics of exhibition making at Te Papa through *Mana Pasifika: Celebrating Pacific cultures* (1997–2007), the Pacific curatorial team's first collective effort to represent Pacific peoples in the reimagined national museum that opened in 1998.

Well before Te Papa formed the curatorial team, the conceptualisation of Pacific-centred exhibition narratives for the new national museum began. This was the starting point for an intervention in the national museum that would radically transform how it presented the national narrative(s) and those of Pacific peoples' cultures and histories. Between 1989 and 1990, the new museum project team organised a series of consultations with stakeholders including Pacific community leaders about the planned exhibitions for the new building. Later, in 1993, Pacific Advisory Committees were set up to progress this work further. Between 1990 and 1992 project staff and museum consultants developed an interpretative plan which included a 'Pacific focus', comprising two exhibitions with the working titles 'Peopling the Pacific' and 'Pacific cultures'. The Peopling the Pacific exhibition would be about 'the story of human conquest of the Pacific, from the earliest beginnings of exploration to the current renaissance of ocean voyaging by Pacific people' and the seafaring and navigational technologies that Pacific peoples used to achieve this (MONZTPT 1994, 2). The unnamed Pacific cultures exhibition would be about Pacific peoples, their societies and cultures and be more comparative and ethnological in its approach.

Early planning documents record that the main reasons for the Pacific focus were the recognition that New Zealand is a Pacific nation, the fact that the museum has a large collection of Pacific Island artefacts, and also the fact that people from the Pacific Island countries are a large, fast growing and visible group in New Zealand, with strong and distinct cultural identities. As Davidson stated: 'But because we are a museum, it is primarily the artefacts that justify a separate Pacific Cultures exhibition'.[2]

We eventually gave the proposed Peopling of the Pacific exhibition the title *Vaka Moana: The peopling of the Pacific* and the Pacific cultures exhibition *Mana Pasifika: Celebrating Pacific cultures.* The concept and initial design work on *Vaka Moana* were well advanced when, surprisingly, museum management reassigned the exhibition space to conference and function purposes to help generate revenue for the museum. The Pacific exhibition team were devastated at this decision. We were left with one allocated exhibition space in the new museum and faced with the choice of continuing to develop *Vaka Moana* or discarding the completed work and developing the proposed Pacific cultures gallery. It is my recollection that Davidson opted for the latter arguing that it would represent a larger number of communities than *Vaka Moana* which, while potentially spectacular, was more narrowly themed and featured far fewer Pacific Islands groups.

From the outset, square metres mattered in the new Te Papa. Exhibition making is often a resource-hungry endeavour. Exhibitions can occupy hundreds, sometimes thousands of square metres of floor space – in other words, the most valuable and contested real estate within museum buildings. Marxist theorist Henri Lefebvre in his reflections on the politics of space argues that space 'is not a scientific object that can be removed from ideology and politics; it has always been political and strategic' (1976, 31). In the heightened creative environment of Te Papa's project offices every square metre of gallery space was subject to strategy and politics. With the cancellation of *Vaka Moana* (1,200 square metres) it became apparent to me that we would struggle for our representation in the national museum with *Mana Pasifika* (279 square metres). The decision felt like an undermining of Te Papa's project to establish our roles as contributors to the project. It seemed that Te Papa in its newly constructed physical presence had yet to escape the ideological space occupied by its predecessor institutions.

The bricks and mortar were fresh, but there was an air of ambivalence in the new Te Papa; despite planning documents acknowledging that 'New Zealand is a Pacific nation' it seemed that there was residue from the old institution. The ideological space of the new museum had already been inhabited and dominated and had not managed to reimagine a post-colonial future. How would it be refreshed? How would the occupation of our new space at Te Papa, already contested on many fronts, be shaped by the institution's ambitions and transformed by our presence?

In creating a new space for Pacific peoples at Te Papa, the physical transformation began with us as new staff of Pacific descent, and our conceptual and literal occupation of the exhibition spaces. We were immediately aware of the necessity to establish authority and credibility in our work process and assure the quality of the content we generated for the exhibition. We utilised several approaches to share authority and manage the politics among ourselves and with our stakeholders. However, for managers above us in the museum hierarchy, staff recruitment was a priority in establishing authority in the overall project. As Pacific exhibition team members we would undertake the research and develop the exhibitions in collaboration with other specialists such as designers and writers in the museum. Our specific roles '… would require not only research skills but also experience and knowledge of one or more Pacific Island cultures, to enable participation in community liaison and consultation, and participation in evaluation exercises' (MONZTPT 1994, 2). We would be the face of the museum in Pacific community interactions and advocates for communities inside the workings of the museum.

The museum management approved the recruitment of the team facilitated by Janet Davidson, a noted scholar and archaeologist with field work experience in New Zealand and other Pacific Islands. In her role as Conceptual Leader, Davidson's academic credentials were important for the establishment of credibility in the new museum project, and its claims to speak with authority. Davidson's history of relationships with Pacific communities and her networks with other museum and university-based scholars saw her well positioned to lead our work.

In the present day, non-indigenous people curating, writing about us and working with us and running institutions in colonised places are points of debate and tension. In this formative period of the Te Papa project, these questions of authority were only just emerging in New Zealand's museums, although Māori had contested them quietly in plain sight for decades (McCarthy 2007). At the time, Davidson had recently transitioned out of her role as the leading voice on Māori ethnology within the museum and been replaced by Māori staff, although they regularly consulted with her on archaeological issues. To my knowledge, questions about Davidson (as a non-indigenous person) leading the Pacific team were not openly voiced among the Pacific community, in part because Davidson had strong relationships in these networks.

I was one of two appointments to the exhibition team that were recruited from an internship programme developing Māori and Pacific staff for the new museum project. My secondment was as an object researcher and collection manager and my fellow intern Fulimalo Pereira was seconded from her curator role at the Auckland Museum project. The appointment of Grace Hutton of Cook Islander/Welsh descent as Exhibitions Interpreter ensured Pacific input into the development of exhibition texts and interpretive media such as mechanical and computer interactives. The (Irish) Sāmoan (me), Tokelauan (Fulimalo) and Welsh/Cook Islander (Grace) makeup of the team could be seen as a strength, bringing some degree of cultural diversity to the project. Our shared training in anthropology brought opportunities and set limitations on our work. On the one hand it meant that we were equipped with tertiary qualifications that are recognised in the museum world. We could draw on a shared vocabulary and some disciplined ways of thinking about, analysing and organising cultural information and materials. On the other hand, our degrees highlighted that we were part of a maligned scholarly discipline among indigenous peoples, including Pacific communities – one associated with colonialism and privileged white male scholars.

Arguably, the makeup of the core Pacific team brought cultural nuance, scientific validity and authority to our work. There were high expectations from museum management and some members of the community. According to museum records, as early as December 1989, at a Pacific exhibitions planning meeting involving museum professionals and Pacific community members, the need was identified to 'have Pacific Island staff. Pacific Islanders are the best people to promote Pacific Island cultures, to welcome people and help other staff' (MONZTPT 1989, 53). Personally, at times I found this authority, this burden of representation, difficult to carry and deliver on. While as a team we had confidence as researchers and museum professionals, being asked what the Pacific view on X, Y or Z is, as if we were a walking Wikipedia on the region, made one's eyes roll. This is a set of conditions we continue to work under until this day and is similarly experienced by Māori curatorial staff in the museum (Cairns 2020).

However, the flipside of this situation is that we gained a small measure of autonomy and non-interference within the institutional processes that other disciplines did not enjoy. I observed that curatorial areas such as Art and Science had a wider threshold for popular/non-specialist engagement, everyone felt they had some knowledge and cultural competency and could have a say in the development of the exhibits. And of course, almost everyone was entitled to contribute, but only a few would be held accountable within the terms of their job descriptions. Curators had to set a direction, negotiate all these incoming opinions, and keep their projects moving forward. In the early 1990s, inside and outside the museum, the academic fields of Pacific history and cultures were more specialised, with a smaller pool of academically qualified staff and people with first-hand cultural knowledge and a cross-cultural, historical understanding of the region. Therefore, the circle of influence and potential intervention in our curatorial work was smaller and the processes of negotiation more economical, but no less fraught.

As anthropologist Kirin Narayan has remarked 'A "native" anthropologist is assumed to be an insider who will forward an authentic point of view to the anthropological community', or in our case, the museum. She goes on to highlight how '... the notion of the "native" anthropologist as carrying a stamp of authenticity is particularly obvious in the ways in which identities are doled out to non-Western, minority, or mixed anthropologists so that exotic difference overshadows commonalities or complexities' (Narayan 1993, 676–7). For the Pacific team as new staff, museum management may have assumed our authority, cultural competence and knowledge just because we were of Pacific Islands descent. We had a kind of insider status similar to, but greater than, participants and observers. One might say we were born to the 'culture' and then university educated. In reality, we were also outsiders by association with the museum and our discipline. These attributes did not gain us many rights and privileges within our communities. I personally had very little cultural authority outside the museum.

Mana Pasifika celebrated cultural survival. It was about the persistence of Pacific peoples and their cultural values and practices in New Zealand. But for us, politically it symbolised our survival in the institutional setting of the museum. Conceptually, we attempted to rework the approach of its predecessor *The Hall of Pacific Peoples* at the former Dominion Museum site. However, rather than organise the displays by island group, we named exhibition segments after established cultural and ethnographic categories such as ceremonies, religion, costume and regalia, food and feasting, music, sports and warfare. Within each of them, we highlighted examples of material culture from various island groups to represent differences and similarities between cultures as well as current and past cultural practices. Many of the objects were sourced from or related to the Pacific cultures with the largest populations in New Zealand.

If this sounds like a representational nightmare, it was. How do you represent the Pacific, comprising so many histories, experiences, peoples, languages and cultures, in one exhibition in a very limited space? Even with our focus on specific objects and their stories it seemed an insurmountable task. Of course, you are drawing from a collection of around 17,000 objects; it might sound counter-intuitive, but even this large number of objects has its limitations. As Pacific Studies scholar Teresia Teaiwa (2015) once said, 'You can't paint the Pacific with just one brush stroke', capturing well the burden of attempting to represent the Pacific in any exhibition, publication, documentary or media, but also challenging the idea that in the first place we can attempt to generalise and represent it too easily.

The conceptual scope and representational challenge of *Mana Pasifika* materialised in the centrepiece of the exhibition in a coloured drawing by artist Michel Tuffery that was reproduced as a large mural. His brief was to synthesise the conceptual elements of the exhibition into one image that would operate as a visual drawcard for people approaching the space from a distance. In Te Papa management-speak of the time, this aspect of the exhibition was described as the 'wow' factor, and every exhibition had to have one. Tuffery was qualified for this task of fusing all the visual and conceptual elements of *Mana Pasifika*. He was established in his art practice, familiar with the material culture of the Pacific region and possessed a visual style and vocabulary to synthesise these elements in an accessible way. His high public profile as an artist meant that he brought his own ethnographically informed authority to this representational task. However, when we visited him while he was undertaking the work, he expressed to us the difficulty he was experiencing incorporating all the exhibition's themes and representing the seven major island groups in one image. It was paralysing. He said it was one of the most difficult pieces he had ever attempted. His challenge also proved a challenge for the curating of the exhibition, with Tuffery's composition of the artwork directly mirroring our curatorial assembling of objects from different historical and cultural contexts, and geographical regions, and the stories to go with them (see Figures 6.1 and 6.2).

Another mechanism we used to draw attention to the cultural diversity of Pacific peoples was the use of indigenous greetings signage in the languages of the seven largest Pacific Island populations in New Zealand. They were placed on a prominent wall and were visible from the entry to the exhibition. This may seem a blunt and obvious approach, but, nonetheless, we deemed it a crucial one for highlighting linguistic and cultural diversity. As scholars of language have noted in other specific institutional and ideological contexts, in an exhibition space, and an ideological space such as a museum, there is tremendous scope for different meanings and social and political effects to be actioned through language or any other linguistic form (Gal 1995, 419). The 'welcome sign' proposal was suggested by Pacific Advisory Committee members who felt it was 'an example of [their] ownership', and referred to the Pacific Island peoples rather than the institution of the museum (MONZTPT 1996b, 1). They commented that the individuality of the

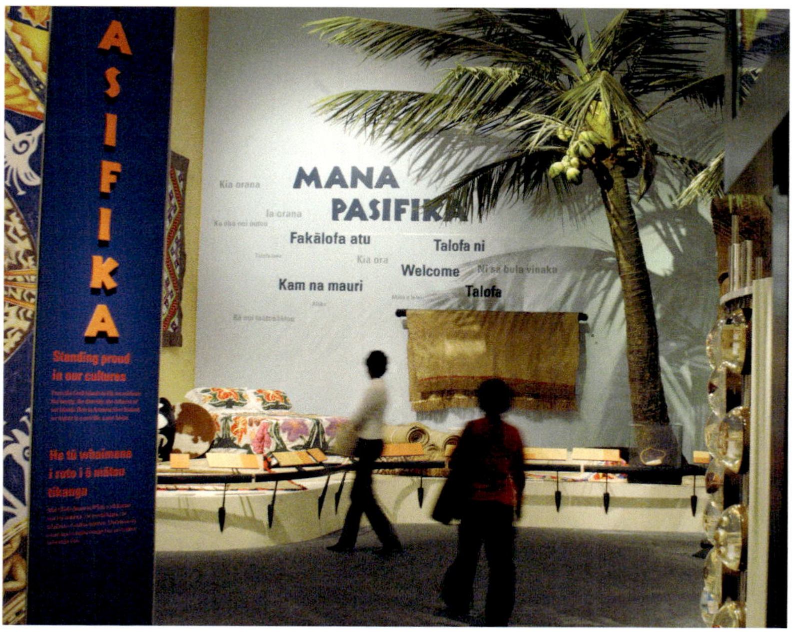

Figure 6.1 *Mana Pasifika: Celebrating Pacific cultures* exhibition.
Source: Museum of New Zealand Te Papa Tongarewa.

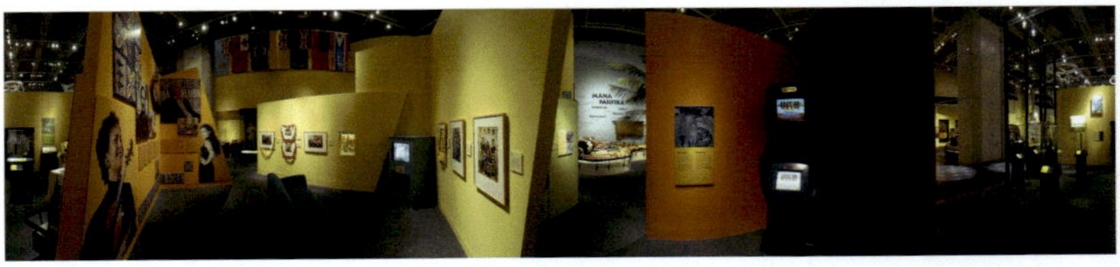

Figure 6.2 *Mana Pasifika: Celebrating Pacific cultures* exhibition [panorama]. Source: Museum of New Zealand Te Papa Tongarewa.

different Pacific nations was important to retain (MONZTPT 1996a, 1) and would help to 'provide a welcoming ambience for Pacific Island visitors' (MONZTPT 1996c, 89). This concern with heterogeneity was also evident in the way committee members gently expressed their concern about the dominance of Sāmoan culture and language in New Zealand, often as a proxy for all Pacific cultures.

Indigenous languages were also used in exhibition label texts. They were a key device in giving visibility to the heterogeneous nature of Pacific peoples in New Zealand and internationally. Our cultural advisors argued for bilingual texts (English and Pacific Island languages) and said it was essential to use the correct word for the language concerned, for example, the many different terms for tapa/bark cloth (MONZTPT 1994, 5).

Another intervention in label text-writing came from Albert Wendt, a Sāmoan author and Professor of English at the University of Auckland and an influential member of one of the Pacific Advisory Committees. He suggested that we avoided italicising indigenous terms, and that where possible we privileged indigenous language over other English translations or glosses. As ethnographers know, translating cultural concepts or names of objects in concise terms is not a word-count-friendly exercise. Where possible we named objects in the language used in its place of origin, if known. For feature objects, we devised short descriptions relating to function and context and translated them into the languages of the culture of origin.

Activating Mana Taonga through indigenous languages provided a mechanism to expand our general visitors' cultural vocabulary and respect those visitors whose cultural artefacts we were putting on display. This may appear a small adjustment in our practice, but it was a significant building block in recentring Pacific people and decolonising the museum's displays. As Tuhiwai Smith argues, 'By "naming the world" people name their realities. For communities there are realities which can only be found, as self-evident concepts, in the Indigenous language; they can never be captured by another language' (1999, 179).

I have written elsewhere (Mallon 2010) about how our experience developing *Mana Pasifika* (and other exhibitions), sensitised us to working with and around visual tropes and stereotypes in our displays. In the exhibition development meetings for a later exhibition called *Tangata o le Moana: The story of Pacific people in New Zealand* (2007–present), we jokingly banned the vocalisation of descriptors such as 'vibrant', 'colourful' and 'exotic', although they continued to find their way into planning documents. Of course, these are often accurate descriptions of the island environments of Pacific peoples and their cultural activities, but in a museum they can become limiting and problematic representations.

An example in *Mana Pasifika*, regrettable in hindsight, was our use of colour in the overall spatial and visual design (Mallon 2010). From the outset, there was a desire to create 'the Pacific' in the exhibition space we were allocated. Our exhibition goals emphasised the importance of colour and a graphic designer came up with a Pacific colour palette of oranges, yellows, greens and, of course, an ocean blue. Unfortunately, the effect of this colour scheme on our exhibition space was to drown the objects in a blown-out background of an overwhelmingly 'hot orange' colour field. A great blue wall (sky or ocean) featured our aforementioned Pacific greetings in woodcut lettering. On reflection, the colour scheme probably reveals our internalisation of the visual cues of the Pacific and how we overemphasised their importance in the museum environment. We had stepped back to the diorama-style displays of the past. Here, the attempt was not to recreate a Pacific village scene or place objects 'in situ', but rather to locate the whole exhibition space 'in situ' – in the Pacific, with 'orange' sunshine, blue water, and the odd piece of corrugated iron as an urban marker.

Our attempt at reconstituting a Pacific environment is part of a museum exhibition phenomenon that for visitors 'feeds the hunger for difference, recreating the travel experience at one remove' (Pieterse 1997, 166). Unsurprisingly, our blue sky in *Mana Pasifika* was a trope that appeared also in major Pacific exhibitions in the United States in the decades preceding Te Papa. Anthropologist Miriam Kahn, in a critique published in 1995, describes how in the Margaret Mead Hall in the American Museum of Natural History 'a permanently blue sky, bereft of rain or clouds, oppresses the visitor' and how that permanently blue sky in the Field Museum Pacific exhibitions, serves as a backdrop to a sandy coral beach, in an exhibit where there are no Pacific peoples, but instead 'just visitors tromping through their idyllic, but vacated spaces' (Kahn 1995, 334).

In retrospect, it seems we could not even untangle *ourselves* from our own institution's long established 'traditions' of ethnographic museum display. In one sense or another, all of us were involved in the kind of interpretive anthropology that Clifford outlines as a reading of culture and society as an assemblage of texts, in our case interpreted from our experiences in the field, of 'our lives', and our anthropology training.

In the process of making this exhibition and engaging in this correspondence, we were party to these decisions. We consulted with a small group of Pacific designers about some of our decisions; we were influenced by Pacific Advisory Committee members who wanted to avoid 'drab cases and dull walls', and the museum project management insistence that a key way to get people into the exhibition was to make it colourful and exotic. One only needs to review the approved exhibition development documents to see how often the word 'colour' appears as part of our curatorial goals. Management suggested other exotic alternatives to create a 'wow' factor, such as 'extra palm trees', and one senior staff member suggested we could have 'a large flat of say Cook Islands church interior altar wall' and present the communities with '... a gathering space ... for religious services – space for 100 people'. Another proposed that the 'sound of a choir could ring out from time to time ... It could also act as a draw card to the Pacific area as visitors investigate where it is coming from' (MONZTPT 1995; 1996a, 1). These suggestions were ignored by the curatorial team, not only for logistical and design reasons but also because they promoted the stereotype of all Pacific peoples being church-going, hymn-singing Christians.

Some of the shortcomings in our collective cultural knowledge and authority were mitigated by the formation of Pacific Advisory Committees, who aided us in managing the cultural expectations and the requirement for authority in our exhibition work. We convened two advisory committees as part of the development process for the two new planned exhibitions, *Vaka Moana: The peopling of the Pacific* and *Mana Pasifika: Celebrating Pacific cultures*. One committee consisted of indigenous and non-indigenous academics, with specialists in education, linguistics, anthropology, history and literature. The other committee comprised a group of indigenous community leaders, academics and artists. We also met with a third committee of design consultants to advise on initial design ideas.

We asked the committees for commentary on exhibition concepts and guidance on issues ranging from cultural expertise to language and customs, access to community networks and other experts, and questions of what subjects and themes should be emphasised in the exhibition. One theme that troubled the curatorial team related to cannibalism and the inclusion of nineteenth-century cannibal forks from Fiji. We were worried about the potential of upsetting cultural sensitivities and compromising the cultural safety of staff. Would visitors be horrified or offended? Albert Wendt put the issue into perspective for us, insisting we include the cannibal forks, put them in the exhibition but also said 'just don't make a big deal out of it, don't sensationalise it. We [Pacific peoples] were cannibals!'.

Similarly, the advisory committee asked us not to overemphasise sports in the exhibition, although at the time it was an area of New Zealand's cultural life where Pacific Islanders were enjoying success and a very high public profile. This was in an era where mainstream media were noticing the 'Polynesianisation of New Zealand sport' (Hyde 1993) and Pacific men and, to a lesser extent, Pacific women, were becoming the international face of professional televised sports such as Rugby Union, Rugby League and Netball. However, while our people were excelling at sports, the advisory committee did not want us to reinforce the stereotype that sport was all we were good at – that we were all physicality and brawn. They urged us to highlight Pacific people's valuable and intellectual contributions to the arts, education, politics and other facets of society.

To address the stereotype of the Polynesian sportsperson we devised a display of large photographic portraits of Pacific Islanders' role models or success stories. They were part of a staged living room setting in the exhibition. This portrait wall of role model individuals faced another wall of family group portraits adorned with synthetic flower leis where we highlighted the achievements of families who had migrated to New Zealand to build new lives and seek educational and economic opportunities for themselves. It was an exercise in establishing authority through polyvocality, utilising a host of personalities and experiences to offer a sense of a connected but diverse community. However, despite our display solution and the mandate of our advisory committees, not all Pacific Islanders were impressed with this approach. The advisory committees who helped us in the development of the exhibition concepts could not shield us from critique. Other cultural authorities out there voiced their displeasure with our curatorial choices, expanding the polyvocal dimension of our work.

The late Jim Vivieaere, an influential Cook Islands artist and curator, was critical of the achievers' wall. He queried if we even needed it at all, sardonically referring to it as 'the hero wall'. However, even if he would not care to admit it, he was one of our heroes, with his ground-breaking art practice and curatorial work, so much so that he had already accepted our invitation to contribute a modest installation work to *Mana Pasifika*. In a more public (published) review of Te Papa's opening exhibitions, and one of the rare examples that addressed the *Mana Pasifika* exhibition specifically, Denis Dutton, a New Zealand academic and philosopher of art expressed his dismay at our 'Hero wall', recounting to his readers how 'Overbearing photographic blow-ups of Pacific Islanders and videos celebrate Pacific identity in the same space that trivialises Polynesian art' (1998, 23). Dutton was seemingly unaware that high-achieving role models such as the ones we displayed were in demand in New Zealand's Pacific communities at this time.

Educational institutions and government agencies profiled Pacific success stories in many areas of endeavour as examples for young people to encourage them to pursue tertiary study and other career training opportunities (Pacific Island Employment Development Board 1988–2005). A motivation behind our celebration of Pacific peoples' achievement in the sporting arenas and other areas of New Zealand's cultural life was the contrast it provided to the very public racism and ambivalence Pacific peoples experienced in the previous decades, particularly during dawn raids in the 1970s where police indiscriminately targeted Pacific peoples in the hunt for immigrant overstayers (Anae 2012). The 'Hero wall' was a representation with an intention of empowerment through centring stories of self-determination and success.

Dutton's disappointment with our trivialising of Polynesian art highlights another shift in our representation of Pacific peoples at Te Papa at that time. Our display approach often accounted for the artefacts' aesthetic and formal qualities; this was well established in the museum's history of display. Indeed, in many cases objects were in the museum for their exceptional formal qualities. However, we also chose to emphasise their social and cultural histories, and their use and form in contemporary contexts. Turning his attention to other displays in *Mana Pasifika*, Dutton exclaimed:

> Most offensive of all, a deeply patinated 19th-century kava bowl from Tonga is forced to share its glass case with a plastic ice cream container. All over the world, marvellous indigenous carving and pottery traditions have died, thanks to the importation of cheap aluminium and plastic containers. This is hardly something to celebrate, and that aged bowl deserves the dignity of an attention undistracted by its tacky, modern surrogate. (1998, 23)

However, *Mana Pasifika* was not the only opening exhibition to come under scrutiny for such unsightly inclusions and confronting juxtapositions. For example, in the art exhibitions, Dutton was horrified to see a revered large-scale painting by celebrated New Zealand artist Colin McCahon displayed next to everyday household items. He wrote:

> In this motley confusion, you'll find Colin McCahon's 'Northland Panels', one of this country's most significant post-war paintings, jostled by an old TV and a Toby jug on one side, and some Hamada pottery and a 1959 Kelvinator Foodarama fridge, complete with display from a department store window ... (1998, 23)

Leading New Zealand-based Sāmoan choreographer Lemi Ponifasio was also unimpressed by our approach to display in *Mana Pasifika*, particularly the family group portraits adorned with synthetic flower leis, not to mention the plastic flower arrangements in a reconstruction of a contemporary altar from a Cook Islands church. He expressed this to me personally and was critical of the aesthetics of the objects and the overall display space rather than the content or messaging. In an interview published in November 2000, a journalist asked Ponifasio, 'What would you like to see at Te Papa's Pacifika [sic] exhibit in 50 years' time?' He replied:

> I think we have to remove some of the crap in there. We are still interested in this 'culture at risk'. My culture is not at risk. Samoans have always been there. But what is presented in the Pacific arts [exhibition] is ... you know, plastic flowers. I hate all that. It doesn't represent me. I don't have those things in my house. It's silly, it's patronising. (Ponifasio in Tucker 2000, 25)

If our anthropologically informed and lived experience as participant observers secured the 'scientific validity' of having these particular objects on display, they did not represent how some or all of the Pacific peoples would choose to be represented within the display culture of the museum. Other Pacific staff in the museum shared Ponifasio's view and questioned some of the curatorial selections of objects. One example was a pair of white 20-litre plastic buckets used by butchers in New Zealand to store and transport what Sāmoans call *povi masima* (salted meat). In Sāmoan communities, *povi masima* is presented at ceremonies such as weddings as part of gifting customs, alongside other cultural valuables such as *'ie tōga* (fine mats). One of my colleagues criticised my intention to include them in the display, saying it was like showing the world our dirty laundry. It was as if the ceremonial presentation of salted meat was somehow something private or a practice that brings embarrassment to the community (seen on any weekend in Sāmoa or New Zealand at weddings or funerals, the more *povi masima* one presents – sometimes entire sides of cattle – the more prestige it brings upon one's family). It was as if seeing ordinary, everyday, utilitarian things like plastic buckets or plastic flower leis in an exhibition representing the cultural lives of Pacific peoples was somehow incongruous within a museum. It seems that the ethnographic reality of contemporary Pacific community cultural life didn't align with some of our stakeholders' ideas of what a museum should display. It is a situation that brings to mind a particular *Far Side*™ cartoon by Gary Larson depicting anthropologists approaching a native's home, while the natives rush around inside panicking and hiding their television sets, electric lamps and VCR players, yelling 'Anthropologists! Anthropologists!' in warning to each other. I can imagine a similar scenario where they are looking out of their window, see us coming and start yelling 'Pacific curators! Pacific curators!'.

Several developments highlighted the thinking of our stakeholders and their agency within the power dynamics of their relationship with Te Papa. There were occasions when the people and communities that we worked with tested the power and structure of the museum and its authority as a cultural institution. Our engagements with the committees were dialogical in nature, a form of discourse where what we produced in the exhibition was in a polyphonic/multivocal and open-ended mode, one that would continue to refresh the representation of Pacific peoples in the museum for years to come. A couple of examples stand out for the way our communities made us stop and reconsider what we were doing.

As I have written elsewhere, Albert Wendt in another of his interventions around language in the exhibition, requested that we abandon the use of terms like 'traditional art' in our labels and display signage. 'Traditional means nothing to me!' he said. At the time, I didn't understand what he was talking about. I had gone through university learning about traditional societies. As a person of Sāmoan descent, I was used to hearing about our traditional culture and customs. As a museum worker, I talked about traditional cultures all the time. However, Wendt was asking us to decolonise the language we use in our exhibitions. In his view, the word 'traditional' as used in categories such as 'traditional arts' and 'traditional practices' was the vocabulary of Western ways of writing about and cataloguing indigenous peoples. We in museums had bought into it and our communities had internalised it (Mallon 2010).

Later, in an interview I had with him in 2008 he explained his position to me in more detail:

> I came to feel very uncomfortable with terms such as traditional, folk history, folk art … Colonial scholars and researchers used them whenever they referred to us but not to their cultures. Such terms I concluded were part and parcel of the Eurocentric colonial vocabulary. Traditional inferred our cultures were/are so tradition-bound they were static and slow to change; that they weren't dynamic and growing and changing; that because they were slow to change and fixed in history they were 'simple and easy to understand'.
>
> Traditional also had implications about how we were viewed as people even to the extent that, because we were tradition bound, we behaved out of habit and past practice and [were] slow to adapt to other ways or change our own ways, that we didn't want to think for ourselves, or were incapable of individual thinking and expression. (Mallon 2010, 368)

Wendt first published these ideas in his inspiring essay 'Towards a New Oceania' (1976). He argued that 'There is no state of cultural purity (or perfect state of cultural goodness)' and warned of stagnation as 'an invitation for a culture to choke in its own bloody odour, juices, and excreta'. He reminded us, 'No culture is ever static and can be preserved … like a stuffed gorilla in a museum' (Wendt 1976, 58, 53, 52).

In his request to our Pacific curatorial team at Te Papa, Wendt was challenging us to decolonise our curatorial practice, and to take care with the concepts and language we used in our representation of Pacific peoples. However, Wendt's questions around tradition were not purely academic or relevant only to museums and the development of ethnographic displays. Pacific artists, painters, sculptors, choreographers, dancers, composers, tattooists and even orators have struggled with the 'traditional' in 'traditional Pacific arts' and the limits it places on their practices (Mallon 2010, 374). Wendt's request around our avoidance of the word 'traditional' persists to this day, despite the occasional transgression.

Te Papa's Pacific Advisory Committees were set up as mechanisms to facilitate Mana Taonga through the advice from and connections to communities they represented. The task of identifying and donating new items for the exhibition not only activated the relationship between the community and the museum, it gave material form to the committee members' knowledge and expertise. The Pacific Advisory Committees were symbolic representations of the community but also of Te Papa's re-orientation towards more collaborative community-informed work. At times, they provided a foil for the curatorial team but also served as a higher cultural authority the team could appeal to. They progressed the development of the museum in museological terms, in theory and practice, and their appointment offered some political protection for Te Papa.

The committees provided members with benefits and opportunities to enhance their social and cultural status within their communities, the university and among their peers. The Pacific Advisory Committees largely represented a cultural elite: people who were recognised for their cultural knowledge, leadership and service. Membership of such groups was politically important to leaders in the growing Pacific communities. The symbolic power of representation and personal involvement in a high-profile public project such as the development of the new national museum was valuable to Pacific peoples when other minority groups were looking for similar exposure. For some members of the Pacific Advisory Committees, their participation connected to the politics of power in their universities, the symbolism of academic capital and peer esteem, and the development of scholars.

Aside from these social and cultural elites there were other consultants who were perhaps less widely recognised in their communities but with varying degrees of investment and involvement in what we were doing. These were the individuals and families connected with key objects on display: the artists, the makers, the donors, the descendants of donors and even the language translators. They all played a part in making the exhibition happen and making manifest the shared authority required to complete the task. However, sometimes our consultative approach and commitment to Mana Taonga backfired. During the installation of the last of the artefacts in the exhibition, we invited one of these consultants to put a significant cultural object that originated in their community into its display case. The invitation was in the spirit of Mana Taonga and intended as a gesture of respect for the object. It took less than five minutes but a few days later we got an invoice for $100. It was a return gesture we hadn't anticipated, and we had to scramble to pay it. Being culturally competent sharers of authority had come with a price tag. It was, seriously, a question of 'hang on, can we afford this?'

Moments like these remind me how there are limits on Mana Taonga, sharing authority, and engaging in a democratic process around exhibition making. It can be expensive and time consuming. One of the regrettable aspects of the Pacific Advisory Committees consultation process was that just as they were coming to terms with what a museum could offer, and appreciating its opportunities and limitations, the time to move on to the next stage in development, building and installing was upon us.

We were unable to fully capitalise on the committee members' newly acquired fluency and understanding of the museum context within the timeline of the project. Initially, we were fortunate to enjoy the regular attendance of academics and leaders at meetings, although over time this commitment diminished owing to the many demands they had on their time and Te Papa's timeframes for completing the work. As opening neared, the dialogue with our committees became intermittent; quick decisions were required and there were too many people contributing to the deliberations. In an environment of shrinking financial and time resources, a democratic exhibition development process became unsustainable. In curator Fulimalo Pereira's view, 'We had a lot of feedback, but along with that you get a lot of different advice on how best to do things ... There was an expectation that there would be a constant dialogue between Te Papa and these communities, but it fell apart really' (Pereira in Williams 1999, 18). Nevertheless, the Pacific Advisory Committees made significant contributions that helped reposition Pacific peoples in the national museum, if not in terms of square metres on the floor, then in our shifted mindsets and our critical awareness around theory and practice in a museum setting. However, as much as the committees had the potential to enhance our curatorial work and safeguard Te Papa from criticism, some members did not hold back from critiquing the museum's exhibition development and consultation processes. About 18 months after opening, in a searing article titled 'Where have all the vaka gone?', Galumalemana Alfred Hunkin, the Chair of the Wellington-based committee was one of three committee members who talked of their disappointment with the outcome of the consultations, and management's decisions minimising the Pacific presence in Te Papa. He said, 'It's the usual Pacific Island mentality. We go along to discuss how good life can be, but we only get a tiny fraction of the pie. We end up saying "let's just do the best we can, because that's how we cope with the disappointment of it all"' (Hunkin interviewed in Williams 1999, 18).

Mana Pasifika was a modest space and a genuine effort to get as much of our collections and stories on display as possible. However, after about 5 or 6 years, it seemed that *Mana Pasifika*, while celebrating our cultural survival, nevertheless persisted in presenting us as exotics.

Part of this may be attributable to the comparative ethnographic approach of the display. We were represented in the museum and somewhat familiar, but not really understood or part of New Zealand society and culture, let alone the grand narratives of Te Papa, the national museum. There was an ambivalence about our presence.

Nevertheless, *Mana Pasifika* sowed the seeds for future thinking about Pacific-related exhibition and collection development. It was a professional development course for the Pacific curatorial team. However, it also had a much deeper impact and legacy, one that has changed the DNA of the museum. Its small presence on the exhibition floor was an intervention in the ideological space, an accountability for the institution based on its own promises around Mana Taonga and its commitments to Pacific peoples. It locked us into the institution's future.

Most importantly, in curating *Mana Pasifika*, we put into action Mana Taonga, and oriented ourselves to collaborative ways of working with the peoples we were representing in our displays. This was an expansion if not a clear leap forward from the work of our institutional and academic predecessors. It was not a process invested in a timeless, ethnographic snapshot approach to display; the project was more open-ended than that. However, I do not want to overstate the positives and argue for some monumental achievement. There were limitations to our realisation of Mana Taonga and this openness, and in fact a narrowness at the time, that has since widened and retracted according to different circumstances. Mana Taonga needs projects to activate it and give it visibility, it is affected by factors such as staff turnover and the loss of institutional knowledge, the tempo of museum programming, and the availability of staff time and funding.

There was an unevenness in the collaborative nature of curating, power sharing and decision making. The museum still made most of the decisions, still held the power, controlled the timelines and, of course, the purse strings. There were restrictions everywhere; inequality in economies of size and scale, square metres, production value, word-counts and even quality of consultation. On reflection, I find it hard to get excited about what we achieved. I felt discomfort and dissatisfaction with the exhibition after opening that lingered with me for a while. My impression is that the impact of *Mana Pasifika* was minimal in the opening weeks and months. Other curatorial disciplines in the museum, which arguably had more cultural capital, more real estate, and financial investment overwhelmed our measly presence. Perhaps it wasn't our time yet to be recognised as ambitiously as we would have liked in such a public space. And could we really demand this level of recognition when the needs and ambitions of Tangata Māori had not been fully realised in the museum nor for the other ethnic minorities in Aotearoa New Zealand?

In the years since opening and by the early 2000s, the presence of Pacific peoples in New Zealand society became even more prominent, especially in highly visible areas of cultural life such as the arts, business, sport and politics. In my mind, it was clear that *Mana Pasifika* was quickly reaching its use-by-date as a long-term exhibition. These circumstances put pressure on Te Papa to rework the existing exhibition *Mana Pasifika* and offer something conceptually fresh. I suspect that *Mana Pasifika* was also an embarrassment to some new people in museum management. There was a very visible disparity between it and every other long-term exhibition on the floor. Some critics were vocal in the media about these issues, others indifferent. A summative evaluation of the exhibition, conducted in November 2001, identified *Mana Pasifika* as one of the most problematic of the Day One exhibitions, with poor visibility and holding power for visitors. By 2005, a business case stated that a new '… *Mana Pasifika* would be one of the cornerstone long-term exhibitions of the Tangata Tiriti quadrant on Level Four' (MONZTPT 2005). A thoroughgoing overhaul of the exhibition was planned. That is another story.

I don't hate ethnography or ethnographic displays, it's a research practice and product that has changed along with anthropology and museums. Curatorial skills and ethnographic approaches to our work are still valuable; what is changing is how we curate, share authorship and processes of developing exhibitions. Our museums will continue to be challenged by the opportunities and trials ethnographic approaches to display present us with. We are having our turn at 'seeing' with greater clarity – our institutions, ourselves and each other – but the politics are morphing into a politics of representation and contestation among peoples who have long been the subject of the ethnographic lens. Today, we talk among ourselves about our growing discomfort curating and initiating projects outside the groups we are culturally connected to. As an indigenous Pacific curator, who am I qualified to represent? Just Sāmoans? Or Irish Sāmoans?

These circumstances lead us towards developing new models of co-curating, co-collecting, collaboration and reciprocation (Mallon 2018). These will inevitably be shaped by the circumstances and locations of where and with whom we work. Over the last decade, we have developed experimental projects of co-collecting and each one is different in design and outcomes. They have included a focus on Master artists in Guahan/Guam, the experiences of Tongan youth in Auckland,

New Zealand (Tonga 2017a; 2017b; 2020), the material culture of German-Samoan colonial legacies (Schorch et al. 2020), Aloha shirts (Regnault 2017), and climate change in the atolls of Tokelau (Yates 2021). We still need to deploy our curatorial toolkits of hard-earned skills and experience, and our knowledge of collections and exhibition making. However, we should do this in a spirit of reciprocity, from the curatorial practice of ethnography as description, to a practice of 'correspondence' (Ingold 2011) where we share with our collaborators the opportunities to see and be seen.

Notes

1 See statement in Te Papa Annual Report 2008–9. Accessed 27 September 2024. https://www.tepapa.govt.nz/assets/76067/1692680461-annual_report_2008-09.pdf, 6.

2 Davidson (c. 1994) Background paper, 'Pasifika exhibitions and collections' (Internal Te Papa document).

References

Anae, M. 2012. 'All power to the people: Overstayers, dawn raids and the Polynesian Panthers'. In *Tangata o le Moana: New Zealand and the people of the Pacific*, edited by S. Mallon, K. Māhina-Tuai and D. Salesa, 221–40. Wellington: Te Papa Press.

Britton, C. 1999. *Edouard Glissant and Postcolonial Theory: Strategies of language and resistance*. Charlottsville: University of Virginia Press.

Cairns, P. 2020. 'Decolonise or Indigenise: Moving towards sovereign spaces and the Māorification of New Zealand museology'. *Te Papa Blog*, 10 February. Accessed 1 September 2024. https://blog.tepapa.govt.nz/2020/02/10/decolonise-or-indigenise-moving-towards-sovereign-spaces-and-the-maorification-of-new-zealand-museology/.

Clifford, J. 1983. 'On ethnographic authority'. *Representations* 2: 118–46.

Clifford, J. and G. E. Marcus (eds). 1986. *Writing Culture: The poetics and politics of ethnography*. Berkeley: University of California Press.

Debary, O. and M. Roustan. 2017. 'A journey to the Musée du quai Branly: The anthropology of a visit'. *Museum Anthropology* 40 (1): 4–17.

Dutton, D. 1998. 'Te Papa: National embarrassment'. *The Weekend Australian*, 6–7 June: 23.

Gal, S. 1995. 'Review: Language and the "Arts of Resistance". Reviewed Work: *Domination and the Arts of Resistance: Hidden Transcripts* by James Scott'. *Cultural Anthropology* 10 (3): 407–24.

Geertz, C. 1988. *Works and Lives: The anthropologist as author.* Redwood City: Stanford University Press.

Hyde, T. 1993. 'White men can't jump: The Polynesianisation of sport'. *Metro*, September: 62–9.

Ingold, T. 2000. *The Perception of the Environment: Essays on livelihood, dwelling, and skill.* London: Routledge.

Ingold, T. 2011. *Being Alive: Essays on movement, knowledge and description*. London: Taylor & Francis.

Ingold, T. 2014. 'That's enough about ethnography!'. *Hau: Journal of Ethnographic Theory* 4 (1): 383–95.

Kahn, M. 1995. 'Heterotopic dissonance in the museum representation of Pacific Island cultures'. *American Anthropologist* 97 (2): 324–38.

Lefebvre, H. 1976. 'Reflections on the politics of space'. Translated by Michael J. Enders. *Antipode* 8 (2): 30–7.

Mallon, S. 2010. 'Against tradition'. *The Contemporary Pacific* 22 (2): 362–81.

Mallon, S. 2016. 'Following Tapa: New contexts and global culture'. In *Made in Oceania: Proceedings of the International Symposium on Social and Cultural Meanings and Presentation of Oceanic Tapa*, edited by P. Mesenhöller and A. Stauffer, 72–86. Newcastle upon Tyne: Cambridge Scholars Publishing.

Mallon, S. 2018. 'Agency and authority: The politics of co-collecting'. In *Curatopia: Museums and the future of curating*, edited by Philipp Schorch and Conal McCarthy, 279–95. Manchester: Manchester University Press.

McCarthy, C. 2007. *Exhibiting Māori: A history of colonial cultures of display*. Wellington: Te Papa Press.

McCarthy, C., E. Dorfman, A. Hakiwai and A. Twomey. 2013. 'Mana Taonga: Connecting communities with New Zealand museums through ancestral Māori culture'. *Museum International* 65 (1–4): 5–15.

MONZTPT (Museum of New Zealand Te Papa Tongarewa). 1989. Pacific Exhibitions Planning, 11 December 1989: 52–3.

MONZTPT (Museum of New Zealand Te Papa Tongarewa). 1994. 'Vaka Moana: The Peopling of the Pacific Concept Development Report', October 1994.

MONZTPT (Museum of New Zealand Te Papa Tongarewa). 1995. Notes of a Mana Pasifika Exhibitions Ideas Session 1.30pm Thursday 21 December 1995.

MONZTPT (Museum of New Zealand Te Papa Tongarewa). 1996a. Mana Pasifika Design Meeting notes 30 January 1996.

MONZTPT (Museum of New Zealand Te Papa Tongarewa). 1996b. Mana Pasifika 100% Concept Design.

MONZTPT (Museum of New Zealand Te Papa Tongarewa). 1996c. Pacific Island Advisory Meeting Notes 22 February 1996.

MONZTPT (Museum of New Zealand Te Papa Tongarewa). 2005. Mana Pasifika Redevelopment Business Case May 2005 (Approval date: 6 July 2005).

Narayan, K. 1993. 'How native is a "native" anthropologist?'. *American Anthropologist* 95 (3): 671–86.

Nencel, L. and P. Pels. 1991. 'Introduction: Critique and the deconstruction of anthropological authority'. In *Constructing Knowledge: Authority and critique in social science*, edited by L. Nencel and P. Pels, 1–21. London: Sage Publications.

Pacific Island Employment Development Board. Annual publication 1988–2005. *Achievers: A guide to scholarships, study grants and awards for Pacific Island students*. Pacific Business Trust; Wellington: Pacific Island Business Development Trust; Pacific Island Employment Development Board.

Pieterse, J. N. 1997. 'Multiculturalism and museums: Discourse about others in the age of globalization'. *Theory, Culture & Society* 14 (4): 123–46.

Regnault, C. 2017. 'Collecting the spirit of Hawai'i through aloha shirts'. *Te Papa Blog*, 12 December. Accessed 9 September 2021. https://blog.tepapa.govt.nz/2017/12/12/collecting-the-spirit-of-hawaii-through-aloha-shirts/.

Schorch, P. and A. Hakiwai. 2014. 'Mana Taonga and the public sphere: A dialogue between Indigenous practice and Western theory'. *International Journal of Cultural Studies* 17 (2): 191–205.

Schorch, P., S. Mallon and N. Tonga. 2020. 'Materialising German-Samoan colonial legacies in refocusing ethnographic museums through oceanic lenses'. In *Refocusing Ethnographic Museums through Oceanic Lenses* by Philipp Schorch, with Noelle M. K. Y. Kahanu, Sean Mallon, Cristián Moreno Pakarati, Mara Mulrooney, Nina Tonga and Ty P. Kāwika Tengan, 121–50. Honolulu: University of Hawaii Press.

Schorch, P., C. McCarthy and A. Hakiwai. 2016. 'Globalizing Māori museology: Reconceptualizing engagement, knowledge, and virtuality through Mana Taonga'. *Museum Anthropology* 39 (1): 48–69.

Teaiwa, T. 2015. 'You can't paint the Pacific with just one brush stroke'. *E-Tangata*, 25 October. Accessed 2 September 2024. https://e-tangata.co.nz/korero/you-cant-paint-the-pacific-with-just-one-brush-stroke/.

Tonga, N. 2017a. 'Kava, fashion, and teenage life: Co-collecting with our Tongan communities'. *Te Papa Blog*, 9 June. Accessed 2 September 2024. https://blog.tepapa.govt.nz/2017/06/09/kava-fashion-and-teenage-life-co-collecting-with-our-tongan-communities/.

Tonga, N. 2017b. 'Countering stereotypes through co-collecting with Tongan youth'. *Te Papa Blog*, 16 June. Accessed 2 September 2024. https://blog.tepapa.govt.nz/2017/06/16/countering-stereotypes-through-co-collecting-with-tongan-youth/.

Tonga, N. 2020. 'Revisiting the Tongan co-collecting stories of *Project 83: Small Things Matter* – Elisapeta Fononga'. *Te Papa Blog*, 7 September. Accessed 2 September 2024. https://blog.tepapa.govt.nz/2020/09/07/revisiting-the-stories-of-project-83-small-things-matter-elisapeta-fononga/.

Tucker, M. 2000. 'Lemi Ponifasio'. *LOOP SWARM* 2000 (Sep/Nov): 25.

Tuhiwai Smith, L. 1999. *Decolonizing Methodologies: Research and Indigenous peoples*. London: Zed Books.

Wendt, Albert. 1976. 'Towards a new Oceania'. *Mana Review* 1 (1): 49–60.

Williams, P. 1999. 'The Lost Vaka'. *LOOP Magazine* 1999 (June): 17–18.

Yates, R. 2021. 'Co-collecting climate change in Tokelau: Project IKA'. *Te Papa Blog*, 3 May. Accessed 2 September 2024. https://blog.tepapa.govt.nz/2021/05/03/co-collecting-climate-change-in-tokelau-project-ika/.

7

The dawn of anthropology and an ethnographic museum in Japan: reconsideration from a post-colonial perspective

Taku Iida

'Decolonise the museum' is a powerful slogan currently used to address the uneven global economic structures resulting from nineteenth-century geopolitics (Soares and Leshchenko 2018; Hicks 2020). However, this may not always be an effective way for certain states to tackle diplomatic or domestic issues. To some extent, this applies to Japan – the focus of this chapter – which has suffered from the colonising pressures of the great global powers in the mid-nineteenth century, but then itself rose to the position of coloniser in the early twentieth century. This ambivalent political status can be observed in the epistemologies and activities prevalent throughout Japanese history. Indeed, some scholars argue that Japan has atoned for its colonial past through US military occupation after its defeat at the end of World War Two in 1945.

Exploring the formation of ethnology in Japan in the 1920s to 1930s, this chapter demonstrates the ambivalent character of this local discipline: colonising on the one hand, and colonised on the other. This does not mean that the colonisers' idea was dominant in one period and that of the colonised was in another; rather, that universal and local aspects of the discipline have been uniquely blended by theories and practices. To demonstrate this, I examine the establishment of the Japanese Society of Ethnology in 1934 (日本民族學會, JSE hereafter) and its impact on the shaping of ethnographic museums in Japan. Although the society is known to have conspired to support wartime imperial policy (Shimizu and Bremen 2003), it also aimed to catch up with Western countries in terms of academic development – that is, Japanese academic imperial tendencies were inspired by both Japanese policy and Western academic trends. As current researchers decide which trajectories to continue exploring, and which to abandon, they need to know what past researchers aimed at in their academic efforts.

Following the section about the establishment of the JSE, this chapter reviews how local aspects of Japanese ethnology developed through two museums. One is JSE's ethnological museum, 日本民族學會附屬民族學博物館,[1] the Hoya Museum, named for the place where it was located. The museum was established in 1937 and opened to the public in 1939, just after the establishment of the JSE, and was closed in 1962. The other museum in focus is the National Museum of Ethnology, Osaka (国立民族学博物館, or Minpaku, to give its public nickname). It was established in 1974 and incorporated the Hoya Museum's collection. The legacy of the Hoya Museum that Minpaku inherited was, I argue, not only the collection but also the strong belief that knowledge of local people is indispensable to building a larger – cosmopolitan or global – society. Tadao Umesao, Minpaku's founder, argued that the Japanese word for museum is not correct because, literally, it means 'house of all sorts of objects'; and that it should instead be a house of all sorts of information, or a 'holotheque' (housing a collection of 'everything' or 'all', as it were). Although this idea is based on a kind of universal science, it clearly denies reductionist ideas which try to summarise details of people's lives into texts, objects or anything else. Experiences of the ethnological/ethnographic museums in Japan will demonstrate their particular context of coloniality and decolonisation, elucidating plural and diverse processes of colonisation.

Prehistory

The year 1884 marks the beginning of modern Japanese anthropology. While interest in different cultures dates back to earlier periods, the study of human groups was not at that time widespread, but was, rather, developed through personal interest. In this regard, the works of famous Japanologists such as Philipp Franz von Siebolt and Edward Sylvester Morse were not closely linked to Japanese knowledge systems. In 1884, Shogoro Tsuboi (坪井正五郎, 1863–1913) established the Anthropological Society of Tōkyō (東京人類學會), which would become the Anthropological Society of Nippon (日本人類學會) in 1941. From a young age, Tsuboi had a passion for organising groups and publishing periodicals, and anthropology was one of his many interests. However, it quickly became his sole focus in life. He became an assistant professor at the Faculty of Science at the University of Tokyo in 1888 and went to Britain to study anthropology from 1889–92. He was promoted to professor in 1892, upon his return to Japan.

In retrospect, Tsuboi's interests and therefore his 'anthropology' covered a much larger area than what we now call anthropology or ethnology, while the present Anthropological Society of Nippon later came to focus on physical and biological anthropology (Hasebe 1939). Tsuboi's wide interests can be compared to the American four-field approach to general anthropology, or to philology in the nineteenth-century European sense, both of which include prehistory. Tsuboi focused not only on all manners and customs, oral traditions and archaeological remains or sites, but also on zoological and palaeontological approaches. He even published works concerning modern folklore or popular culture studies. In downtown Tokyo, he recorded hairstyles, costumes and the footwear of passers-by, and analysed the results in the context of Westernisation. His first book on signboards found in towns also tries to theorise modernisation of the cultural landscape (Kawamura 2013). But these tendencies may have preceded his preference for anthropology. He had been so good at writing Edo-style light poems and essays that his anthropology might have gone further to include cultural studies.

In accordance with Tsuboi's wide interests, the Society consisted of scholars from a variety of disciplines. To give a few examples, Kotora Jimbo, who contributed as a general secretary, was a geologist; Kotaro Shirai, who disputed with Tsuboi on controversial Ainu origin, was a botanist; and Shozo Arisaka, who discovered potteries belonging to the Yayoi tradition, later became an officer in the navy. Apart from Tsuboi, who became an associate professor at the University of Tokyo, no other member earned their livelihood from anthropology.

Tsuboi's broad interests were inherited by Ryuzo Torii (鳥居龍蔵, 1870–1953), known for both his active ethnographic explorations and intensive archaeological excavations. At the age of 16, he became a member of the Anthropological Society of Tōkyō while living in Tokushima, almost 500 km away. In 1892, when Torii was 22, he moved to Tokyo and began to work for Tsuboi. In 1913, when Tsuboi died in Saint Petersburg, Torii was a lecturer at the University of Tokyo. He was promoted to associate professor in 1922 but left the University in 1924. Anthropologists who remained at the University became more engaged in physical or biological anthropology. Although Torii kept teaching in private universities, the official training course of anthropology in the broadest sense had by then disappeared (Torii 2013).

Just as countless buildings were destroyed by the Great Kanto Earthquake of 1923, the field of the wide-scoped anthropology went quiet. This event stimulated university scholars and amateur members of the anthropological societies, to form a new circle around what was then called ethnology in Europe.

Before shifting focus to the 1920s, it is important to mention Tsuboi's engagement in exhibitions with respect to museology. Tsuboi was a universalist. As the first Professor in anthropology in Japan, he went to England and introduced many Western theories, biological or cultural, to Japan. He was not critical of exhibitions of humans as living examples of cultural practices; he even directed a human exhibition at the Fifth National Industrial Fair in Osaka in 1903 (Yamaji 2014), which was to be accused in a later age for having neglected human rights (Matsuda 2003). The cultural 'samples' included Ainu, Ryukyu (Okinawans and surrounding islanders), Taiwanese, Malayan, Javanese, Indian, Turkish and Zanzibari people, among others, while the exhibition of Chinese people was cancelled after its inauguration. The Ryukyu people also withdrew, after arguments led by Ryukyu people living in Honshu, Okinawa and other areas, 'performing' thereafter in the presence of visitors. As Tsuboi had named this pavilion as the 'Scientific House of Man' (学術人類館), he must have intended to introduce Western scientific knowledge to the Japanese general public, as Barth (2011) showed to have been done in Europe and America. In the following year, Tsuboi exhibited ethnographic artefacts at the University using a grid system, supposedly developed for the first time for the 1867 Paris Universal Exposition (Kawamura 2013, 222–7). Tsuboi had the idea of disseminating scientific knowledge among the general public through ethnographic exhibitions.

Although it is not known what artefacts Tsuboi exhibited then, Minpaku received a collection of more than 6,000 objects from the Faculty of Science, University of Tokyo. The Faculty is said to have managed a small gallery of scientific objects which was designed by the zoologist Edward S. Morse. However, Morse's contribution is little known because most of the objects seem to have been collected after Torii's arrival in Tokyo. This collection obviously shows that, after Torii's resignation, ethnographic objects lost their value for mainstream academics. Tsuboi's universalist attitude stood in contrast to his successor Torii, who took more account of reports on local situations than general theories. The JSE prehistory shows us an example of conflict between Tsuboi's universalist and Torii's localist ideas.

Establishment of the Japanese Society of Ethnology

The decade from 1924 (Torii's resignation) to 1934 (establishment of the JSE) was significant for Japanese anthropology for two reasons. The first was the successive foundation of journals targeting both academic and general readers, while the second was the increase of young scholars studying abroad – both of which improved the quality of scholarship and quantity of scholars. From the social viewpoint, this period saw economic stability stimulate mass consumption and the latter promoted academic people's activities in Japan as well as Western Europe.

It is important to recall that the period from 1924 to 1934 saw the advent of radio broadcasting (1925), the popularisation of cinematography (especially due to Shochiku's intensive production after 1926), and the introduction of gravure printing (for example, the Osaka Asahi Newspaper began to print and insert the Asahi Graphic into the newspaper in 1921). Railroads not only expanded urban areas but also increased tourism, while the growth of the socialist movement provided publishers and readers with theoretical works to read. In this context, publishers began to issue periodicals on books and cinemas, collections of classic literature and popular magazines such as *Kaizo* (1919–55), *Kingu* (1924–57) and *Bungei-shunju* (1923–present). In such a circumstance of mass media domination (Iida 2011), Kunio Yanagita (柳田國男, 1875–1962), the founder of Japanese folklore studies (Morse 1990), began to publish *Minzoku* (民族, meaning 'ethnos', 1925–9).

Yanagita began his career as an administrator for agricultural affairs and developed his academic thinking on his business trips. He organised Kyodo-kai (郷土會, literally meaning 'Home Place Association') and issued *Kyodo Kenkyu* (郷土研究, literally meaning 'Study of Home Places', 1913–17), for which he played a role as a general secretary and chief editor under Inazo Nitobe (新渡戸稲造, 1862–1933), who taught colonial policy and applied his own theory when he served as Governor-General of Taiwan. In this circle, aiming at peasants' welfare and societies, Yanagita deepened his ideas of how a society should modernise itself while maintaining its tradition. Therefore, he became more and more interested in the unwritten knowledge of the past, as seen in people's unconscious behaviour. In 1920, when the League of Nations was established, Nitobe was appointed as Deputy Secretary-General. During his tenure, Yanagita also served on its Committee for Mandated Territory in Geneva, from 1921 to 1923. During his time there, Yanagita bought many English, French and German books of ethnology, anthropology, philology and folklore studies.

The first issue of *Minzoku* was published immediately after Yanagita returned from Geneva. This journal was expected, at least by Yanagita, to promote field studies in rural areas, both in Japan and abroad, and to develop this discipline with inspiration from Western theories and ethnographies. Additionally, this journal served as a means to collect little-known information from rural areas, similar to *Notes and Queries* in Britain, and to expand the academic network, as well as to publish articles. Masao Oka (岡正雄, 1898–1982) worked as the chief editor of *Minzoku*. Due to difficulties with Yanagita regarding editing policies, Oka departed to Vienna to study ethnology, supervised by Wilhelm Schmidt (1868–1954), once the final issue was released in 1929. After the JSE was established in 1934, Oka would play a significant role in establishing the nationally managed Ethnos Research Institute (民族研究所, 1943–5), but that is another story (Ishikawa et al. 2016). More importantly, it was Masao Oka's brother, Shigeo Oka (岡茂雄, 1894–1989), who had been Torii's student and subsequently owned the company Oka Shoin (岡書院), which published *Minzoku*.

Another journal *Minzoku-gaku: The Japanese Journal of Folklore* (民俗學, 1929–33) was established to fill the niche left open by the demise of *Minzoku*. For this purpose, academics from diverse disciplines and universities gathered to produce a periodic subscription journal dedicated to the study of ordinary people's lives, whether inside or outside of Japan. The publisher was Minzoku Gakkai (民俗學會, or 'Society of Folklore'), but Oka Shoin also worked as a distributor. Although this society was simply a group of subscribers, the members organised several meetings annually. In addition, the frequency of *Minzoku-gaku* (12 volumes per year) was too high for a purely academic periodical, and *Minzoku* had only appeared six times a year. Both journals were issued on a commercial basis in the 1920s, and ethnology, or folklore studies, was thus beginning to reach a general audience outside academia.

Concerning the second character of the period 1924–34, that is, the increase of young scholars studying abroad, it is noteworthy that the contents of *Minzoku-gaku* were printed in English as well as Japanese. In this period, when the distinction between ethnology and folklore studies (as well as anthropology and philology) was unclear, most academics tried to relate the domestic scene to that in other countries, despite language barriers. Later, Japanese ethnologists would become more interested in anglophone, francophone and germanophone currents than Japanese folklorists, who would focus on Japanese rural areas, paying little attention to foreign academia. This was caused by the ambivalent attitude of Yanagita, who had introduced Western trends but would later distance himself from them. After the demise of *Minzoku-gaku*, Yanagita founded *Minkan Denshō no Kai* (民間伝承の会, literally, 'Society for Popular Traditions'), which would become the Folklore Society of Japan (日本民俗学会) in 1949. The age of *Minzoku* and *Minzoku-gaku* can be summarised as a transitional period when Japanese folklorists were influenced, much more than at present, by foreign academic trends. After the JSE was established, the two disciplines, ethnology and folklore studies, went their own ways.

Communication between Japan and Europe became more frequent in the 1920s, primarily due to three factors (Iida forthcoming (a)). First, transport infrastructure improved dramatically. In 1921, Prince Hirohito had travelled to Europe by way of the British Empire route. On the other hand, the Nihon Yusen Company had built ships for the European route and had strengthened connections between the two regions. Second, security and economy in both regions became stable after World War I and the 1923 Great Earthquake. In Europe, where fierce battles had taken place, both academic and economic activities could once again gather pace. In Japan on the other hand, the new rich were looking for opportunities to increase their capital. There were thus more and more Japanese people trying to encourage their sons and daughters to study abroad. Third, the League of Nations was established in 1920, propagating the idea of replacing force with peaceful diplomacy to prevent another world war. Japanese politicians and diplomats grew more and more interested in cultural exchange with other countries. For example, Eiichi Shibusawa (澁澤榮一, 1840–1931), a businessman known as the founder of capitalism in Japan, got involved in the Doll Exchange Project where more than 10,000 dolls were shipped from the United States to Japan to deepen the friendship between the two countries. In the next decade, the Japanese government would establish Kokusai Bunka Shinko-kai (國際文化振興会, 'Society for Promoting International Culture') in 1934 and merge it with other societies into the Japan Foundation (国際交流基金) in 1947.

Such changes in diplomatic atmosphere unavoidably strengthened the character of academia as a broker of Western knowledge. Kurakichi Shiratori (白鳥庫吉, 1865–1942) and Nenozo Utsushikawa (移川子之藏, 1884–1947), both historians having studied before 1910 in Europe and the United States, respectively, would play important roles in the 1930s and 1940s JSE. Apart from the JSE, Muneyoshi Yanagi (柳宗悦, 1889–1961) rediscovered Japanese beauty in the 1920s Mingei Movement (民藝運動, often summarised as the Japanese Arts and

Crafts Movement) after the introduction of Auguste Rodin's and French impressionists' works to a Japanese audience. Socialist philosophy and Western alpinism were also introduced in this age, as is represented by Eiichiro Ishida (石田英一郎, 1903–1968) and Kinji Imanishi (今西錦司, 1902–1992), respectively, both of whom later became JSE members without studying abroad. Therefore, the JSE was founded in such an atmosphere of a 'Westernisation' of thinking.

As followers of European science, Japanese scholars found themselves in an ambiguous position. Although epistemologically colonised by Europeans, they found themselves politically under Japanese Imperial rule. They relied upon European disciplines of history and ethnology but employed evidence from the new Japanese colonies. In the disciplines of both history and ethnology, scholars applied 'scientific' facts from European academia to the context of new Japanese colonies. Even if history and ethnology dealt with comparatively different materials, that is, the former with written materials and the latter with objects or oral traditions, both appropriated each other's findings to construct the people's past or present mental structure. For this common goal, the two disciplines were inseparable. The first ethnographic museum in Japan was born into this ambivalent position in politics (colonising/colonised) and disciplines (history/ethnology).

The museum circle as an academic new wave

Among the founding members of the JSE, there were mid-career or even young scholars who had studied abroad. Before the JSE's first meeting was held, its aims and plans were announced together with a list of 68 proposers including Enku Uno (宇野圓空, 1885–1949), Chijo Akamatsu (赤松智城, 1886–1960), Takashi Akiba (秋葉隆, 1888–1954), Kosaku Hamada (濱田耕作, 1888–1938), Keizo Shibusawa (澁澤敬三, 1896–1963) and Nobuhiro Matsumoto (松本信廣, 1897–1981). All six scholars had studied in Europe or the United States in the 1920s. Five of the six initial board members also had experience of studying abroad. Compared with this large number of former students in Western countries, Mikinosuke Ishida (石田幹之助,1891–1974) was the only member, surprisingly, who had studied in China. Finally, there were four European names on the list. The introduction of Western knowledge was apparently one of the major concerns for the newly established JSE.

Keizo Shibusawa, successor to both the business and the Viscount title of his grandfather Eiichi, the founder of Japanese capitalism, was by far the most dedicated member in the establishment of the Hoya Museum attached to the JSE. Shibusawa's contribution to academia cannot be overstated. He contributed as scholar, sponsor and an administrator of the JSE at one and the same time. First, as a scholar, he published books and articles on the Japanese fishing industry and its history. He also published historical documents on fishers' lives. For this latter work he was awarded the Japan Prize of Agricultural Science in 1940. He shared with other ethnologists an interest in common people's lives, rather than the elite, although his approach was more sociological or historical. Second, as a sponsor, he supported students studying abroad by providing personal scholarships, as he did for Masao Oka, who had been acquainted with Shibusawa since they were of high-school age, and decided to study in Austria after splitting up from Yanagita. Shibusawa also hired young scholars who pursued topics in his own areas of interest, especially in fisheries and material culture studies. His collection of material culture began to form in the 1910s, and he exhibited it on the second floor of his rickshaw (later automobile) garage. He named his exhibition and the research group 'The Attic Museum'. Later, in the 1930s, he published two monograph series entitled *The Bulletin of the Attic Museum* and *The Attic Museum Notes*, for young authors, whether academic or not. His support covered not only publications but also research trips. After the Hoya Museum and the JSE's Research Institute were established, he donated the land and the buildings, as well as paying the annual salary of the young scholars working there. His annual donation amounted to 1,000 yen in 1940–3, compared to the annual subscription for the JAE as only 10 yen, back in 1835 (Iida 2021).

Third, as an administrator, Shibusawa became a JSE board member (1934–42) and held senior positions in the JSE's succeeding bodies: as Vice-President (1942–5) of the Association of Ethnology, Chairman and President (1945–9) and President (1951–63) of the Japanese Association of Ethnology. For someone who was not formally part of a university, this was an exceptional level of engagement, leaving aside his additional generous financial contribution. In addition, he donated the Attic Museum collection to the Hoya Museum in 1944. When the JSE was established, Shibusawa had been more interested in material culture studies, originally inspired by Western museums, especially Skansen Open Air Museum (Maruyama 2013). From 1928 to 1932, he added many objects from Shitara, Aichi to the Attic Museum's collection. Quite different from most of the objects previously collected, these Shitara objects were not toys appreciated as visual objects since the Edo era, but ethnographic objects that had been produced and used in daily or ritual life in rural areas. Japanese folklorists have called such functional objects *mingu* (folk implements). In this sense, Shibusawa's collection was ethnographic rather than aesthetic. Moreover, his approach was even scientific in anthropological terms.

Shibusawa's collection building was an indirect result of Kotaro Hayakawa's (早川孝太郎, 1889–1956) pioneering monograph on Shitara New Year festivals (Hayakawa 1930). Although this book concentrated on geographic variation and general description of ritual practices and symbolism (and was therefore criticised as lacking religious and socio-economic significance; Shibusawa 1992, 395), it was based on long-term fieldwork, which was common in British social anthropology. Stimulated by Hayakawa's work, Shibusawa shifted the focus of collection from toys to ethnographic objects. Although Shibusawa's interest in socio-economic process derived from his undergraduate thesis on manufacturing industry (Kato 2020), after Hayakawa's publication, Shibusawa became more conscious of ordinary people's absence from the historical record and sought to introduce them through their material culture. This attitude, keeping distance from general history and Western theory as a result, is quite notable among the researchers of his age. Shibusawa's investment in the ethnographic museum should be understood in this context, but not only from a Western viewpoint. The initial impetus for the Japanese museum movement came out of domestic moments rather than colonial.

A brochure of the Hoya Museum issued in 1949 is conserved at Seijo University, as is another version from 1951 at Kanagawa University and Nara National Research Institute for Cultural Properties, while copies are reproduced in publications. The brochure describes the history of the collection but not how it was exhibited, how it was formed or how it was valued as an academic resource. Since exhibitions at Shibusawa's Attic Museum and the Hoya Museum are yet to be explored (see YHM and ISJFC-KU 2002, however), his and his group's publications are good clues to their perspective on material culture studies. To put the conclusion first, although they were not successful in formalising their methodology, it had a functionalist tendency where various dimensions of human life were in play. This tendency is most evident in *Notes and Queries on Folk Implements* (Attic Museum Editions 1937), which comprises descriptions of 120 folk implements with pictures and texts.

The Preliminary Report of the So-Called Ashinaka (Attic Museum 1935; 1936) is another crucial source for understanding Shibusawa and his team's perspective on the study of material culture. Originally published as two articles in the first two volumes of the *Journal of the Japanese Society of Ethnology*, the report focuses on the examination of the Ashinaka, a traditional foot gear crafted from twisted rice straw. The first half of the report is dedicated to analysing samples collected from various regions across Honshu, Shikoku and Kyushu, in order to examine the different methods and techniques used to create and use the Ashinaka. The second half of the report reviews literature, paintings and drawings that depict the Ashinaka, providing insight into Shibusawa's unique approach to understanding the material culture and daily life of past societies. However, it should be noted that the editor of the report states that it remains preliminary and the two approaches are not fully related to each other, and not related to museological approach. The process of this research, including an exhibition as a result, would have been recorded in both photographs and texts, had the visual reproduction technology been more developed. Shibusawa's idea is more suitable in the age of digital technology.

Most of the objects in the Ashinaka book would later be stored at the Hoya Museum and Minpaku. The editor clearly stated that the book aimed to describe each object, which has much variation in itself, as 'not only a mere material existence but also a living process of communication among individuals, households and village societies'. Although it was not easy to document relations between objects and human activities because of insufficient budget and human resources, Shibusawa and his colleagues attempted to give explanations of a folk implement 'through ecological analysis of its raw materials, preparation, production, utilisation, conservation, resolution, disposal, and reduction to materials again, with consideration of people's thought about these stages of process' (Attic Museum Editions 1937, 2–3). As a result, numerous questions and answers were posed, sometimes from a linguistic viewpoint, at other times in relation to symbolic, aesthetic, practical, social, industrial or distributional perspectives. This approach can be called 'object-centred functionalism'. Unlike archaeologists, Shibusawa and his colleagues were conscious of the relevance of material culture studies for contemporary societies and, for this reason, made much of people's thought. This approach is quite unlike Tsuboi's, Torii's or any Western scholar's at that time. Shibusawa was opening up this new field of material culture studies independently.

Refurbishment of the Japanese Ethnographic Museum

In the 1940s, activities at the Hoya Museum stagnated due to wartime economic depression and a split between 'museophilist' and 'bibliophilist' members of the JSE Research Institute emerged. Although both museophilists and bibliophilists were both interested in field methods, the latter segregated themselves into a newly established Ethnos Research Institute and stopped caring for the museum. As a result, the museum was not only obliged to close but was also exposed to risks from air raids by the United States Air Force (Iida 2021). In the 1950s, no progress was made within the discipline of material culture studies, while Shibusawa and colleagues devoted themselves to implementing the 1950 *Law for the Protection of Cultural Properties* (Kikuchi 2001). In the 1954-amended scheme of the Law, the Hoya Museum's collections of backpack frames (背負い運搬具) and of *oshirasama*, or folk religious statues related to silkworm breeding (おしらさま), were designated as Important Folk Materials (later Important Tangible Folk Cultural Properties) which are considered 'indispensable to understand the role and influence of tradition in the daily life of the Japanese'. Prior to this designation, the Hoya Museum had been reopened, in 1952. However, 10 years later, in 1962, it was forced to close again due to Shibusawa's subsequent declining economic circumstances. Fortunately, the collections were donated to the government for future exhibition upon the establishment of a national ethnographic museum, in accordance with Shibusawa's will.

It was Minpaku, established in 1974, which inherited Hoya Museum's missions and collections. The Hoya Museum's objective of disseminating ethnological knowledge among the public was loyally achieved by Minpaku, by both the first Director-General, Tadao Umesao (梅棹忠夫, 1920–2010), and his predecessor Seiichi Izumi (泉靖一, 1915–1970), who were JSE board members responsible for the Minpaku museum's foundation. Its location, on the site of the first World Exposition in Japan in 1970, also had a connection to the Hoya Museum, which had been planned to be nationalised on occasion of the 1940 World Exposition prior to the cancellation of the event. Shibusawa's original intention to demonstrate his collection in a museum related to an unusual event was thus fortuitously achieved. However, based on lessons learned from the Hoya Museum's social isolation in the 1940s, Minpaku stressed its mission to face the future and not the past. It does conserve objects from the past, but Minpaku underlines not its origin but its direction: the future. One of the consequences of this idea is that a museum is not only an exhibiting facility but also a repository for all forms of information: literal, visual, sonic etc. Umesao (1987, 17) argued that the Japanese word corresponding to a museum is not *haku-butsu-kan* (博物館, literally meaning 'house of all sorts of objects') but *haku-joho-kan* (博情報館, 'house of all sorts of information'). He even created the neologism *holothèque* after bibliothèque and discothèque. This concept justifies the fact that there are more than 50 research staff members working at Minpaku on projects which do not directly link to the museum collection.

In 2007, Minpaku launched a new Master Plan for the Exhibitions 2007 which states:

> The situation surrounding ethnographic museums has dramatically changed over the past 30 years. Remarkable advances in the exchange of people, goods and information on a global scale has brought about dramatic changes in cultures of various groups, and accelerated movement of self-reflection of 'our own culture' and 'our own history'. As a result, one-sided ethnographic descriptions and exhibitions are being criticised, increasingly day by day, by the very people who are described or exhibited, or the bearer of the culture. With the dramatic increase of the visitors' knowledge and information, the visitors' demands are rapidly becoming intensified and diversified. In addition, the academic environment has changed dramatically. Regarding the concept of culture, too, we are seeing a great paradigm shift: cultural relativism, dominant 30 years ago, is declining while more dynamic cultural concepts are created, and multiculturalism is rising on the background of increasing clashes and conflicts between cultures. (National Museum of Ethnology 2007, author's translation)

Based on such recognition, Minpaku renovated its galleries from 2008 to 2017 (Yoshida 2017). Although the Master Plan has an objective and neutral atmosphere, Minpaku has hosted repeated discussions of the politics of representation (Takezawa 2003; Kawaguchi and Yoshida 2005), sometimes as a concerned party of critics or disputers (Niessen 1994; Ohtsuka 1996; Shimizu 1996; Pietersma 2023). Such discussions have been more bibliophilic than museophilic in the actual Japanese context, where Euro-American anthropological trends directly affect Japanese ones. This would therefore be of no interest to taxpayers. However, the dichotomy between museophilic and bibliophilic is not productive at all. Shibusawa wished to achieve harmonisation of both, though he did not achieve it. But it could be more feasible if we focus more on museum objects, as Shibusawa did.

In 2014, Minpaku started a series of sub-projects under an umbrella project called 'Info-forum Museum'. In this project, the database of museum objects, usually focusing on a particular area, is regarded as a tool to communicate with local communities. I have myself also been responsible for two info-forum databases of the Africa collection (Iida forthcoming (b)), as my disciplines include African studies. With help from the community, the project members in Osaka reflect local information of each museum object. Even if the objects are stored in Japan, if we can reflect on the people's memory in the database and feed back the museum visitors' ideas to the people, then the objects can keep living with the people. Based on the same perspective, we are also trying to connect the objects with Osaka residents of African origin through museum activities. This collaborative process not only promotes community members' participation in documentation and curation (Clifford et al. 2020), it also facilitates communication between unrelated groups and brings about valuable information to academia. Such academic knowledge will be not only fresh to visitors but also communicative, because visible and tangible objects play a role as a fount of knowledge. Thus, Shibusawa's attempt in *Notes and Queries on Folk Implements* has been re-energised with the help of digital technology, merging the museophilic and bibliophilic, and mediating academic and non-academic at the same time.

Conclusion

Museums are embedded in society. This is the very reason why museums should be decolonised. However, their historical contexts may not necessarily be colonial. The Hoya Museum was rather distant from the 'mental colonisation' which affected most of the JSE members when they attempted to relate their activities with Western academic trends. This comparative independence led the museum to a temporary stagnation in the 1940s to 1960s. However, now that digital technology has facilitated image reproduction and complex hyperlinking of various information, knowledge of ethnographic objects is increasingly shared with remote areas. Consequently, museum objects are regaining their original contexts, even while they retain the context of the museum where they are stored. This change is profitable from a museophilic perspective, wherein people from any background can learn from objects independently of Western trends. Museums are being rapidly decolonised with the help of digital technology.[2]

Notes

1 To be precise, the JSE changed its name in 1942 and again in 1945, becoming first the Association of Ethnology and then the Japanese Association of Ethnology.

2 I wish to acknowledge that this research was supported by JSPS KAKENHI Grant Number JP19H01400 (2019–22) and Minpaku's info-forum database project 'Building a multilingual and interactive database for the Africa collection' (2017–22).

References

Attic Museum. 1935. '所謂足半（あしなか）に就いて [On the so-called Ashinaka, a kind of Japanese Straw Sandal: A preliminary report]'. *Japanese Journal of Ethnography* 1 (4): 710–68b. https://doi.org/10.14890/minkennewseries.1.4_710.

Attic Museum. 1936. '所謂足半（あしなか）に就いて [On the so-called Ashinaka, a kind of Japanese Straw Sandal: A preliminary report (concluded)]'. *Japanese Journal of Ethnography* 2 (1): 115–245. https://doi.org/10.14890/minkennewseries.2.1_115.

Attic Museum Editions. 1937. '民具問答集 [Notes and queries on folk implements]'. Tokyo: Attic Museum.

Barth, Volker. 2011. 'Des hommes exotiques dans les expositions universelles et internationales (1851–1937) [Exotic men in worldwide and international exhibitions 1851–1937]'. In *Exhibitions: L'invention du sauvage* [Exhibitions: The invention of the savage], edited by P. Blanchard, G. Boëtsch, and N. J. Snoep, 180–205. Paris: Actes Sud et Musée du Quai Branly.

Clifford, James, Atsunori Ito, Reiko Saito, Kenji Yoshida, Isao Hayashi and Taku Iida. 2020. 'International symposium "Future of the Museum: An Anthropological Perspective"'. *Bulletin of the National Museum of Ethnology* 45 (1): 115–76.

Hasebe, Kotondo. 1939. '人類學雑誌の前身 [Predecessor of the Japanese Journal of Anthropology]'. *Japanese Journal of Anthropology* 54 (8): 53–5.

Hayakawa, Kotaro. 1930. '花祭 [Flower Festival]'. Tokyo: Oka Shoin.

Hicks, Dan. 2020. *The Brutish Museums: The Benin Bronzes, colonial violence and cultural restitution*. London: Pluto Press.

Iida, Taku. 2011. '日本人類学と視覚的マスメディア——大衆アカデミズムにみる民族誌的断片 [Japanese anthropology and visual mass media: Ethnography in popular academism]'. In 山路勝彦編『日本の人類学——植民地主義、異文化研究、学術調査の歴史 [Japanese Anthropology: History of colonialism, cross-cultural studies and academic research], edited by Katsuhiko Yamaji, 611–70. Nishinomiya: Kwansei Gakuin University Press.

Iida, Taku. 2021. '財団法人日本民族学協会（1942年～1946年）と附属博物館（1937年～1962年）——アーカイブズ資料をとおしてその性格をふり返る [Japanese Society of Ethnology (1942–1946) and its Ethnological Museum (1937–1962): Reconsideration of its character with reference to archival materials]'. *Journal of Cultural Anthropology* 85 (2): 336–46.

Iida, Taku. Forthcoming (a). '両大戦間期の日本民族学——フランスとの関係を中心に [Japanese ethnology in the inter-war period with special reference to relations with France]'. In ファシズム期のナショナリズムとインテリジェンスに関する人類学史 [Anthropological History of Nationalism and Intelligence in the Fascist Era], edited by Katsumi Nakao. Tokyo: Fukyosha Publishing.

Iida, Taku. Forthcoming (b). 'Making museum objects revive in human societies: An experience of an image sharing project between Africa and East Asia'. In *Cultural Transmission Against Collective Amnesia*, edited by Taku Iida. Osaka: National Museum of Ethnology.

Ishikawa, Hideshi, Josef Kreiner, Ken'ichi, Sasaki and Takehiko Yoshimura (eds). 2016. *Origins of Oka Masao's Anthropological Scholarship: Proceedings of the International Symposium at Meiji University, November 27, 2015*. Bonn: Bier'sche Verlag Anstalt.

Kato, Koji. 2020. 渋沢敬三とアチック·ミューゼアムーー知の共鳴が創り上げた人文学の理想郷 [Keizo Shibusawa and the Attic Museum: A humanities utopia created by the resonance of ideas]. Tokyo: Benseisha Publishing.

Kawaguchi, Yukiya and Kenji Yoshida. 2005. *Representing African Art and Cultures*. Osaka: National Museum of Ethnology.

Kawamura, Nobuhide. 2013. 坪井正五郎ーー日本で最初の人類学者 [Tsuboi Shogoro: The first Japanese anthropologist]. Tokyo: Koubundou.

Kikuchi Akira. 2001. 柳田国男と民俗学の近代ーー奥能登のアエノコトの二十世紀 [Kunio Yanagita and the Modern Age of Folklore: Twentieth century of Aenokoto of Ishikawa Prefecture]. Tokyo: Yoshikawa Kōbunkan.

Maruyama, Yasuaki. 2013. 渋沢敬三と今和次郎ーー博物館的想像力の近代 [Keizo Shibusawa and Wajiro Kon: The modernity of the museographical imagination]. Tokyo: Seikyusha.

Matsuda, Kyoko. 2003. 帝国の視線ーー博覧会と異文化表象 [The Imperial Gaze: Expos and representations of the cultural other]. Tokyo: Yoshikawa Kōbunkan.

Morse, Ronald A. 1990. *Yanagita Kunio and the Folklore Movement: The search for Japan's national character and distinctiveness*. New York: Gerald.

National Museum of Ethnology. 2007. 国立民族学博物館における展示基本構想 2007 [General Exhibition Plan 2007]. Suita: National Museum of Ethnology. Accessed 28 September 2024. https://www.minpaku.ac.jp/sites/default/files/museum/exhibition/main/main_basicconsept.pdf.

Niessen, Sandra A. 1994. 'The Ainu in Minpaku: A representation of Japan's indigenous people at the National Museum of Ethnology'. *Museum Anthropology* 18 (3): 18–25.

Ohtsuka, Kazuyoshi. 1996. 'Exhibiting Ainu culture at Minpaku: A reply to Sandra A. Niessen'. *Museum Anthropology* 20 (3): 108–19.

Pietersma, Edwin. 2023. 'From crafts to agency: The legacy of colonial discourses in exhibiting the Ainu in the Tokyo National Museum and National Museum of Ethnology at Osaka between 1977 and 2017'. *Museum and Society* 21 (3): 22–35.

Shibusawa, Keizo. 1992. 澁澤敬三著作集 第3巻 [The Collected Works of Keizo Shibusawa, Vol. 3]. Tokyo: Heibonsha.

Shimizu, Akitoshi. 1996. 'Cooperation, not domination: A rejoinder to Niessen on the Ainu Exhibition at Minpaku'. *Museum Anthropology* 20 (3): 120–31.

Shimizu, Akitoshi and Jan van Bremen. 2003. *Wartime Japanese Anthropology in Asia and the Pacific* (Senri Ethnological Studies 65). Osaka: National Museum of Ethnology.

Soares, Bruno Brulon and Anna Leshchenko. 2018. 'Museology in colonial contexts: A call for decolonisation of museum theory'. *ICOFOM Study Series* 46: 61–79. https://doi.org/10.4000/iss.895.

Takezawa, Shoichiro. 2003. '民族誌博物館の現在ーー民族誌博物館は21世紀に存在しうるか? [Ethnographic museums today: Can they exist in the 21st century?]'. *Bulletin of the National Museum of Ethnology* 28 (2): 173–222.

Torii, Ryuzo. 2013. ある老学徒の手 [Memoirs of an Old Scholar]. Iwanami Shoten.

Umesao, Tadao. 1987. メディアとしての博物館 [The Museum as a Medium]. Tokyo: Heibonsha.

Yamaji, Katsuhiko. 2014. 大阪、賑わいの日々ーー二つの万国博覧会の解剖学 [Osaka and its Lively Old Days: Anatomy of the two world expositions]. Nishinomiya: Kwansei Gakuin University Press.

YHM (Yokohama History Museum) and ISJFC-KU (Institute for the Study of Jomin Folk Culture, Kanagawa University) (eds). 2002. 屋根裏の博物館ーー実業家渋沢敬三が育てた民の学問 [Attic Museum: People's science enhanced by businessman Keizo Shibusawa]. Yokohama: Yokohama History Museum.

Yoshida, Kenji. 2017. 'Minpaku: Forty years since opening, and what lies ahead'. *Minpaku Anthropology Newsletter* 44: 1–5.

8
Redefining ethnographic museums and ethnographic displays in China: a century-long debate

Luo Pan

In 2013, scholars from Europe and North America met at a conference at Oxford University to discuss the dilemmas and possible futures facing museums of ethnology. They agreed that an ethnographic museum was no longer a place for the public to experience something new but had become instead a space for constant reflection on colonial history, thanks to public pressure. They also agreed that ethnographic displays, which were used to showcase 'cultural others', remained the foundation of ethnological museums and that they still function to explore a world full of differences (Zetterstrom-Sharp 2013, 27).

However, the focus of Chinese anthropologists with regard to ethnographic museums in the twenty-first century is different. From the 1920s to the present, Chinese scholars have continued to debate the definition of 'museum of ethnology' in China. They constantly ask, what should a museum that showcases the various ethnic groups in China be called? The core of these debates is about what anthropologists can do to represent a country like China, with its many ethnic groups, through museum exhibitions. These unresolved debates reflect the changing relationship between the discipline of anthropology and anthropological museums in China over a period in history when Chinese anthropologists participated in the governance of relations among ethnic groups. The Chinese debates demonstrate the relevance of Chinese anthropologists' museum-related practices to anthropological theories, marking the beginning of the development of a Chinese anthropology, and its localisation (人类学中国化) in the study of Chinese civilisation. The debates represent the Chinese anthropologists' efforts to interpret the nation as a whole through museum exhibitions. These discussions indicate that Chinese anthropologists and politicians have not reached a consensus on how to portray a 'united country with multiple ethnic groups'. This remains true today.

Systematic studies of museums in China have been a quite recent effort. Scholars including Tracey Lu (2014), Xu Jian (2016) and Duan Yong (2017) worked on the origin and development of modern museums in China. Some have cited the ethnographic collections that anthropologists made, but the considerable efforts to build a national ethnographic museum have got short shrift due to the different research foci among scholars. Others like Kirk A. Denton (2013) and Marzia Varutti (2014) have argued that museums are conducive to the consolidation of Chinese nationalism. However, without serious and sufficient observation, 'ethnic museums' are simply tools that 'serve the purpose of making visible the political authority and the glorious narratives of the Chinese State where these most need to be enforced' (Varutti 2014, 268), and ethnic theme exhibitions are 'the unilateral vision of a group of actors – Chinese public authorities – mediated by the museum' (Varutti 2014, 270). Kirk A. Denton has discussed the existence of a National Museum of Ethnology in China (Denton 2013, 215), but his study has limitations.

What is the attitude of anthropologists to a national museum dedicated to ethnic minorities in China? In this chapter I describe the development of ethnographic museums (museums of ethnology) in China from the perspective of the disciplinary history of anthropology and the role of Chinese anthropologists in leading the development of a unique system of 'museum anthropology'.

1920s to 1940s: museums of anthropology and ethnological collections

William C. Sturtevant points out that in Europe and North America after the 1920s the relationship between 'academic anthropology' and anthropological museums was interrupted due to the differences in their research interests (Sturtevant 1969, 625). However, it was around this time that Chinese anthropologists began to establish a Chinese national museum of anthropology. This was a period of rapid theoretical advancement in the history of anthropology featuring the debates between evolutionists and diffusionists. China was then encountering foreign aggression and there was a deep sense of crisis over national sovereignty. Against this backdrop, the idea of a National Museum of Anthropology was proposed, initially as a civic educational institution which could support research. The purpose of this museum, amid the changing domestic and international environments, has constantly shifted. Its collections and exhibitions showed a transition from evolutionary theory to diffusionism, and its educational mission gradually developed from popularising ethnography to imparting the concept of a united China. As a result, such a museum was gradually expected to take on the responsibility of promoting mutual understanding among ethnic groups and ethnic unity and in forming a new sense of community through exhibitions.

On 14 February 1921, Cai Yuanpei (蔡元培)[1], the first Minister of National Education in the Republic of China and President of Peking University, gave a public speech titled 'What is culture?' (Cai 1921, 10–15). In this speech he set out for the first time his vision of an educational system in China which included a Museum of Anthropology (人类学博物馆) to represent different ethnic groups. Educated at the University of Leipzig, Cai followed the European model and proposed a 'museum of history' with displays of ancient relics to show the Han people's history of civilisation, and a 'museum of anthropology' to display daily utensils, clothes, decorations, models of houses as well as pictures of customs from all the other minorities, to educate the public about the 'differences between the civilised and the barbarian wild' ('文野之辩') (Cai 1921, 14). In 1928 when he headed the Ethnology Group of the newly founded Academia Sinica, Cai started to contact a large number of young anthropologists who had returned to China after studying abroad. He launched a large-scale survey and started to make a collection of ethnic cultural relics.

As China's external crises intensified, museums were given greater political significance, and ethnic-themed exhibitions attracted great attention due to their important role in constructing national unity. Exhibitions showcasing China's ethnic groups demonstrated the multiple influences of evolutionary theory and diffusionism, which are also closely tied to national unity and a republican ideology. An example of this was the *Models of Chinese and Foreign Nationalities* exhibition ('中外民族模型展') held by the Henan Museum of Nationalities in October 1928 (see Niu 2017). The Henan Museum of Nationalities was formerly the Henan Museum, which was renamed under the direction of Feng Yuxiang (冯玉祥), the warlord in power in Henan, to promote Sun Yat-sen's Three Principles of the People (*Sanmin Zhuyi* 三民主义), nationalism, democracy and people's livelihood, and to highlight the unity of China. The *Models of Chinese and Foreign Nationalities* exhibition aimed to 'demonstrate the process of human evolution, inspire invention and creation, spark revolutionary thought, and encourage revolutionary spirit by displaying the process of transformation and evolution' (4).[2] After Feng left Henan in 1929, the name of the museum was changed back to Henan Museum. After the 1930s, the museum still focused on ethnic exhibits, and its curators continued to emphasise that 'all museums should make contributions to the Chinese nation' (Wang 1936, 1).

In 1933, Cai Yuanpei urged the Ministry of Education to establish a preparatory office for a 'Central Museum' in Nanjing, with three departments for humanities, crafts and natural sciences. The establishment of the Central Museum was closely related to the ongoing discussions over China's frontier governance and the composition of Chinese people in the context of the war against Japan. It was also a great collective attempt by scholars in the Republican Era, mainly those who had studied abroad, to learn from the West how to build museums in China. Specifically, the Department of Humanities aimed to 'promote scientific research and support popular education' (21),[2] with the goal of 'enlightening the Chinese people on the evolution of Chinese ethnic groups and culture and promoting ethnic consciousness' (30).[2] The collections displayed in this department were 'largely related to ethnology, anthropology, archaeology, all about the evolution of human culture' (4).[2] These interpretations all show distinct evolutionary methods. However, the Central Museum had an interdisciplinary structure, in line with Cai's early proposal of a 'five-category museum' structure. The purpose of integrating nature, history, humanities and crafts was to boost mutual understanding among various ethnic groups (Cai 1921, 10–15).

Historical materials and artefacts from border areas were the two most important types of collections in the Department of Humanities. In collecting ethnic cultural relics, the field approach adopted by the Ethnology Group of the Academia Sinica in the late 1920s continued. Ethnic cultural relics were preserved as collections and as part of the national-level museum. This practice comprehensively redefined the academic value of ethnic cultural relics in museums and, more importantly, the political significance of these relics as museum collections.

In this context, anthropologists' collections of ethnic artefacts and the related exhibitions represented the idea of 'a nation [that] includes numerous ethnic groups' in nation-building. The overall purpose of the Department of Humanities was to combine archaeology, history and ethnology in order to present a full picture of China's history of ethnic integration and the cultural interaction between ethnic groups, providing historical and ethnological references for the cultural unity of the Chinese nation to meet the needs to shape a new culture featuring ethnic diversity and national unity. A consensus was reached in Chinese academia from the 1920s to the late 1940s to combine the perspectives of cultural similarities and to acknowledge diversity based on unity (Liu 2018).

Similar ideas for constructing a national-level museum showcasing ethnic artefacts was proposed by the anthropologist Tao Yunkui (陶云逵) in 1941. He proposed a National Borderland Humanities Museum, with the headquarters located in the capital and four branches in Lanzhou, Chengdu, Kunming and Guilin. Tao envisioned the museum as an instrument to enhance mutual understanding and interactions among ethnic groups, in order to 'cultivate national consciousness' and 'enable the people to have a correct understanding of borderland culture' (Tao 1941/2017, 206). His plan for collecting and displaying ethnic artefacts specifically noted that it is necessary to observe the 'response of an ethnic group to different cultures' (Tao 1941/2017, 207). Additionally, the locations of the museum symbolised the relationship between the Han 'Huaxia' (华夏) and the 'Four non-Han peoples' (四夷). Tao's ideas represented the objectives of anthropologists of the time to strengthen the sense of identity in the Chinese nation. However, with his premature death, the plan failed to go further.

In 1944, the collecting of ethnic artefacts grew rapidly, and the Central Museum held several ethnic-themed exhibitions to promote understanding among the Chinese people (Xu 2016, 140). In 1948, the Department of Humanities of the Central Museum was completed in Nanjing. On 29 May in the same year, the museum joined the National Palace Museum in Peking (Beijing) to hold an exhibition. The collections from the National Palace Museum were mainly calligraphy works and paintings from various dynasties and porcelain from the Song and Qing dynasties (960–1368 CE), while the Central Museum displayed bronzes from the Shang and Zhou dynasties (1600–256 BCE), cultural relics from the Han dynasty, imperial portraits from various dynasties, and ethnic artefacts from various ethnic groups in southwestern and southern China, portraying 'an outline of Chinese culture' (Zeng 1956/2009a, 93–4). The juxtaposition of archaeological relics of the Central Plains Dynasty (中原王朝) and ethnographic artefacts of southwestern ethnic groups in the Department of Humanities was a political metaphor to explain the connection between southwestern ethnic groups and Chinese culture as a whole. It not only represented a change in the philosophy of collection and the methods adopted but also suggested the inclusion of ethnic minorities as an integral part of the whole Chinese culture in the national ideology.

1950s: de-anthropologisation and the formation of a new narrative

In the 1950s, after the founding of the People's Republic of China (PRC), the Central Museum was renamed Nanjing Museum. The central government proposed to build a national museum in Beijing to showcase the minority groups within its borders, which sparked new discussions in Chinese society, not only among anthropologists. The key issues included the scope of the collection of ethnic artefacts in the new period, how to collect them, the content of exhibitions and the name of the museum. As a result, scholars and the central government agreed on the name the Museum of Nationalities (民族博物馆) instead of the Museum of Anthropology/Ethnology (人类学/民族学博物馆), to underscore the unity of Han Chinese and other ethnic groups. However, it was called the Cultural Palace of Nationalities (民族文化宫), rather than the Museum of Nationalities, the title that had been previously agreed upon. The display at the Cultural Palace focused on ethnic policies rather than the history and cultures of ethnic groups. Methodologically and theoretically speaking, the new museum followed a path towards de-anthropologising the representation of ethnicities. The exhibitions held by the state set the basic framework for the subsequent ethnic narrative in museums.

After 1950, the proposal to build a national-level museum to showcase ethnic minorities was put back on the agenda due to several factors. First, after 1950, the central government frequently sent delegations to visit minority areas and invited representatives from these areas to visit Beijing. These interactions resulted in the accumulation of a large number of gifts from these regions (Ma 1999) which awaited cataloguing. Second, a series of field surveys and social history investigations around 1950 helped collect ethnic artefacts. Under the initiative of anthropologists like Wu Zelin (吴泽霖) and others, many ethnic cultural relic exhibitions were held across the country. The scale of the collection of artefacts and materials was large enough to build a museum. Third, building museums of nationalities in ethnic minority areas was also an important way to promote ethnic policies. In August 1950, the National Ethnic Affairs Commission began to consider establishing a museum, and the Preparatory Office for the Central Museum of Nationalities, headed by Wu Zelin, was established. In the same year, the Ministry of Culture held a symposium to discuss the scope of collecting ethnic cultural relics.[3] On 19 August, Zeng Zhaoyu (曾昭燏), given her experience in holding ethnic cultural relic exhibitions for the 1948 exhibition at the Central Museum, was commissioned by the Preparatory Office for the Central Museum of Nationalities to draft 'The scope of collecting cultural relics from ethnic groups in China' (Zeng 1956/2009b, hereafter referred to as 'The Scope'). The Scope suggested that 'the collection should focus primarily on each ethnic group's unique features, attention should be paid to things that display significant invention or artistic value, and efforts should be taken to show the lifestyles, competencies and knowledge of ethnic groups' (Zeng 1956/2009b, 62). The Scope proposal, though somewhat brief, included collections reflecting social class, national heroes and the history of the revolutionary struggle. The Scope to some extent continued the way anthropologists had grouped artefacts by cultural areas under the influence of diffusionism in the 1940s. This approach recognised cultural exchanges between regions and continued the consideration of the unity of Chinese culture. The same train of thought was also found in archaeological work. For example, Chen Mengjia (陈梦家) noted in 1954 that 'we can see from the cultural relics that our Chinese nation … has such a long history, covers such a large area, has many ethnic groups, and can be seen in thousands of ancient cultural relics, but with an unquestionable commonality' (Chen 1954, 67). In 1956 Zeng Zhaoyu again approved of collecting items by cultural areas in her revision of the Twelve-Year Vision Planning for the Nanjing Museum (Zeng 1956/2009b, 75–85).

In 1955, Premier Zhou Enlai (周恩来) mentioned to the anthropologist Yang Kun (杨堃) in Kunming that 'museums of ethnology in other countries only reflect the backward aspects of ethnic groups in dependent colonial states, but we need to promote the social, historical, and cultural aspects of various ethnic groups, their influences on us and their neighbors, and their contributions to the creation of a great country like China' (Ma 1988, 12).Then, in 1956, the anthropologist Fei Xiaotong (费孝通) made a speech proposing the establishment of a national museum. Reflecting on how anthropology was a colonial discipline, Fei pointed out that most Western museums of ethnology display only those identified as 'the barbarian other'. Hence, he suggested:

> We should not follow the Capitalist route, where they study their own culture as folklore and other ethnic groups as ethnology. Therefore, I object to label those museums that collect, study and

display Han people's lives as 'museums of folklore' and those that display ethnic groups as 'museums of ethnology'. In my opinion, these two types should be collectively called museum of nationalities (民族博物馆). (Fei 1956/1988, 109)

Fei proposed to use the term 'museum of nationalities' to create a new type of museum, different from the European tradition of museums of ethnology. Undoubtedly, such a museum would have a symbolic meaning in political terms, and Fei hoped that it would fuse political and academic functions and present both Han and other ethnic groups as one. In the same year, anthropologists including Yang Chengzhi (杨成志), Wu Wenzao (吴文藻), and Pan Guangdan (潘光旦) drafted the Twelve-Year Plan for the National Museum of Nationalities, suggesting that anthropologists and the central government had reached a consensus to build a National Museum of Nationalities different from 'museums of ethnology in other countries' (Ma 1988, 12).

However, in 1957, the National Ethnic Affairs Commission officially proposed a Cultural Palace of Nationalities (CPN) to celebrate the 10th National Day in 1959. The project was approved in August of that year, incorporating a museum as part of it. The CPN had been proposed by Chairman Mao Zedong (毛泽东) at a meeting of the Political Bureau of the CPC Central Committee in 1951. Mao envisioned a cultural palace for ethnic minorities which could symbolise the unity of various ethnic groups (Zhao 2006, 7). In September 1958, while the design of the CPN was underway, the State Council defined it as 'an exhibition hall for historical relics of ethnic minorities in which economic and cultural achievements of ethnic minority areas are frequently displayed' (Liu 1999, 22).

The most important part of the National Day celebration was a CPN exhibition that opened in 1959. Unlike all previous collections, artefacts and relics used in the 1959 exhibition were collected by provinces under orders from the central government. The first collection was made in November 1958, when the State Council issued the Notice on Collecting Exhibits and Books needed for the Cultural Palace of Nationalities ('The Notice'). The Notice specified three main tasks for museums and libraries: publicising the achievements of the party in the work on ethnic affairs, reflecting the achievements made by ethnic minorities in history and their contributions to the country, and conducting education on the unity of all ethnic groups to build the country. The Notice listed eight categories of exhibits to be collected: six were related to China's achievements in the work on ethnic affairs, and two were about 'historical artefacts' and 'everyday utensils'. Under its direction, about 50,000 cultural relics were identified and collected for the CPN. The second phase of collection took place on 21 March 1959, when the Ministry of Culture issued the Opinions on the Allocation of Ethnic Cultural Relics from the Palace Museum.[4] The Cultural Palace of Nationalities, History Museum and Palace Museum were invited to discuss the matter, and decided to relocate all the ethnic cultural relics in the Palace Museum that were collected after the founding of the PRC to the CPN. In addition, all the gifts that the ethnic minorities offered to the central government during the early days of China were assigned to the Palace. This relocation provided the CPN with sufficient materials and exhibits for the 1959 exhibition to fully demonstrate the central government's control over the territory and the implementation of ethnic policies. The overall significance of the CPN was established.

The 1959 exhibition was the outcome of the cooperation by different provinces. A draft outline for the exhibition was sent to different provinces along with The Notice. The second article of the outline stated that 'China has been a unified multi-ethnic country since ancient times', and the remaining sections focused on the achievements made in the work on ethnic affairs after the founding of the PRC (Fan 2016, 1–17; 2008, 97–108). All provinces were asked to participate in discussions, develop proposals and submit materials and exhibition checklists in a unified format for reference. Some provinces clearly stated that the focus of the exhibition should be on the present rather than the past to prevent a display of cultural relics, while other provinces included content related to the origin and history of ethnic groups. Due to the different understandings among provinces, how to narrate history quickly became the most debated issue in the development of the outline.

In January 1959, Sa Kongliao (萨空了), the Deputy Director of the National Ethnic Affairs Commission, gave a speech entitled 'On the Cultural Palace of Nationalities' at the directors' meeting of the commission. He clearly defined the historical part of the exhibition as the history of 'revolution and struggle' and stated that the focus of the exhibition should be on the achievements made in the 10 years since the founding of the PRC (Sa 1959b). The commission organised a conference on the topic of the museum from 31 March to 10 April 1959 for the National Day celebration. Sa once again proposed that:

> Our goal is to use history to demonstrate that China has been a unified multi-ethnic country since ancient times, in order to consolidate national unity, further enhance the solidarity among

> all ethnic groups, and conduct internationalism and patriotism education. Special emphasis should be placed on the new atmosphere of socialist ethnic relations that has been formed. (Sa 1959a)

The exhibition's theme was *Achievements in the Work on Ethnic Affairs Ten Years after the Founding of the PRC* (hereafter the *Achievements Exhibition*),[5] and the timeline was set from the founding of the nation to 1959. The promotional leaflet and related reports emphasised that the exhibition was indeed based on the principle of 'focusing on the present rather than the past'. 'A unified multi-ethnic country since ancient times' and the history of revolution and struggle were mentioned briefly, while socialist transformation was the focus of the narrative (Su 1959, 42–7). The exhibition covered three halls of the CPN. The 'central hall' displayed the achievements made in the work on ethnic affairs across the country. The 'regional hall' was divided into 15 sections, reflecting the construction of various ethnic autonomous regions, and the 'special hall' exhibited the implementation of ethnic policies in provinces which were not populated by ethnic minorities (非民族聚居省份).

The *Achievements Exhibition* was on display for seven years as an ongoing exhibition of the CPN and played a significant role in publicity work. Its content and framework had a particularly far-reaching influence on the subsequent local museums of nationalities. For a long time after 1959, the museum under the CPN served as a national-level museum of nationalities. In October 1979, the museum under the CPN held a new *National Exhibition of the Work on Ethnic Affairs* to celebrate the 30th anniversary of the founding of the PRC, which retained the ideas, form, and structure of the 1959 exhibition.[6] These two exhibitions at the CPN exhibited the basic requirements for a national level 'museum of nationalities' at that time.

Although the *Achievements Exhibition* did not display the history of a 'unified multi-ethnic country' or that of 'revolution and struggle', these two themes were taken up as important content in exhibitions held by various museums of nationalities. The Five Series of Books on Ethnic Issues, compiled from 1958 to 1964, provided a reference for these narratives.[7] These historical records, which were researched and compiled according to a unified plan, had similar narrative patterns. The first step was to trace the origin and historical development of an ethnic minority and make judgements on the social and historical stage (such as slave society or feudal society) of the ethnic group. The second step was to briefly describe the contributions that the ethnic group had made to 'building a great unified country' in history and culture, such as technological or artistic achievements. The third step was to describe how the ethnic group became a part of the Chinese nation at a certain historical stage, often with heroic activities in the revolutionary history as a clue. The final step was to present the political, economic, and cultural development of the ethnic group after ethnic identification. This four-step narrative thread has been widely used in ethnic-themed exhibitions frameworks, where emphasis has been put on 'us' at different levels. As a result, 'the narrative forms of the Chinese nation and various ethnic groups are like a genealogical tree in terms of logical significance and genre' (Fan 2008, 97–108).

During this period, anthropologists had suggested the new concept of 'a museum of nationalities'. However, the discipline of anthropology had been demonised as a bourgeois social science, intertwined with European colonialism in the 1950s. During the Cultural Revolution it had been abolished in colleges and universities completely and was now isolated, shunned by the CPN along with the idea of museums of nationalities. The CPN was instead responsible for collecting ethnic cultural relics, and its exhibitions were completely focused on ethnic policy and publicity work. However, the *Achievements Exhibition* was also withdrawn in 1967 because of revisionist policies.

In the mid-1980s, the CPN began to hold commercial exhibitions, and its role as a cultural and commercial space outperformed that of a museum. After the economic reforms of the period, international interest in cultural preservation made it urgent for China to comprehensively promote the protection of ethnic cultural relics and cultural heritage. At this time, the CPN could no longer fulfil the responsibilities of a national museum of nationalities.

1980s: towards comprehensive museums of nationalities

In the early 1980s, the discipline of anthropology was gradually revived in universities and colleges, and there were also calls for a new national museum to display and preserve ethnic culture. Yet the question remained whether this museum should be named a 'museum of nationalities' or a 'museum of ethnology'. During this period, returning anthropologists proposed to establish an ethnological system with Chinese characteristics (Liang 1980, 12), and confidently believed that the new system could academically support the museums of nationalities. Most of them agreed upon the name 'museum of nationalities' and that such museums should integrate historical, archaeological and anthropological methods. From the 1980s to the twenty-first century, museums of nationalities in China focused on the collection of ethnic cultural relics and sought approval from the central government to build a physical museum. A considerable number of anthropologists were involved.

In the 1980s, there were two main reasons for the call to rebuild a national museum of nationalities. First, the Cultural Relics Protection Law (文物保护法) was issued in 1983, with the concept of 'cultural heritage' gaining a growing influence. This development enhanced the protection of ethnic cultural relics and significantly expanded the scope of the collection of these relics. In this context, it was necessary to build museums of nationalities in ethnic regions. Second, the revival of ethnology and anthropology in Chinese universities provided the academic support for the building of museums of nationalities. The First National Conference of Ethnology was held in Guiyang in October 1980. This was attended by 233 anthropologists and ethnic culture workers. They discussed the name of the museums, their research objectives and development plans for the discipline of ethnology, in which many regarded museums to play an integral part (Chen 1980; Liang 1980; Jin 1980).

In February 1983, Hu Qiaomu (胡乔木), then member of the CPC Politburo, argued that the museum industry needed a gradual but significant development plan and stressed that 'there were currently no museums of nationalities' (Hu 1984, 7). In March, the National Ethnic Affairs Commission invited the United Front Work Department (统战部); the Ministry of Culture (文化部) (current name Ministry of Culture and Tourism); a group of experts including anthropologists Yang Chengzhi, Song Shuhua (宋蜀华) and Wu Zelin; and Zhang Bo (张镈), designer of the CPN, to meet for a discussion and then to submit a proposal to the State Council to establish a national museum of nationalities. This proposal received a reply from State Council in April stating that 'museums would not be established for only academic purposes'. Ma Yin (马寅), who was in charge of the preparatory work for the museum of nationalities, also emphasised in a reply that the Museum was 'not for academic purposes'. On 26 June the National Planning Commission (国家计委) approved the construction of the museum in the latter part of the Seventh Five-Year Plan (七五计划).

In 1984, the preparatory group organised the National Conference on Cultural Relics Work, and the establishment of the preparatory group for the Chinese National Museum of Nationalities was announced at the conference (Wu 1984). The museum was accorded 'immeasurable significance' including 'reproducing the history of all ethnic groups', 'reflecting the great leap of all ethnic groups after the founding of the PRC', and 'promoting national confidence and national self-esteem'.[8] At the same time, the museum was expected to contribute to the tourism industry. Starting in 1985, Ma Yin (马寅), planned a diverse knowledge training scheme for members of the preparatory group, covering ethnology, archaeology, ethnic policies and theories, studies of museums of nationalities, history of material culture and photography of cultural relics (Yin 2007, 191). Historians Song Zhaolin (宋兆麟) from the History Museum and anthropologists including Wu Zelin, Lin Yaohua (林耀华), and Yang Chengzhi participated in the training. The preparatory group also contacted universities such as Peking University, hoping that they would provide staff from their museum and anthropology departments. The construction plan received extensive attention from domestic and foreign cultural heritage organisations such as UNESCO. It also attracted the attention of anthropologists from around the world (Ma 1987).

Historians, archaeologists and anthropologists participated in discussions during the preparatory stages of the museum, and there was constant debate over the museum's name and the direction of collecting ethnic cultural relics. The most intense discussions took place from 1981 to 1985. There were, broadly, three views on the museum's title. Most scholars did not specifically draw a clear line between the two names, 'museum of nationalities' (Minzu Museum) and 'museum of ethnology' (Minzuxue Museum). They believed that ethnology underpins museums of nationalities. Ethnologist Liang Zhaotao (梁钊韬), for example, combed through the methods of Chinese-style ethnic investigations and identification and proposed developing an ethnological system with Chinese characteristics based on the absorption of 'diffusionism' and 'Marxist anthropology' and building a central museum of nationalities (Liang 1980, 18). In Liang's plan, the 'museums of nationalities' had the function of promoting the traditional culture of the ethnic groups and the ethnic policies (Liang 1980, 18). In 1981, Lin Yaohua described the function of this kind of museum (whether museums of nationalities or museums of ethnology) as not only exhibiting, studying, and preserving cultural relics but also promoting policies formulated by the party and the country to the general public, displaying research findings and enhancing ethnic solidarity (Lin 2014, 55). Yang Kun used both 'museum of ethnology' and 'museum of nationalities' in his article. More specifically, the former was used as an approach, and the latter as the museum's name, which received recognition from the staff members of the museum

(Yang 1986). Some scholars believed that museums of ethnology should bear the responsibility for disseminating national ethnic policies. For instance, Chen Kejin (陈克进) suggested that museums of nationalities be renamed museums of ethnology because the former, in terms of item collection, identification, research and its social function, cannot develop to the same level as the latter, or its research achievements (Chen 1997, 109). Wang Zhaowu proposed a comprehensive museum of ethnology in the capital city of Beijing, which would represent all ethnic groups across the country and showcase their history (including the Han) (Wang 1982, 283–7). In the meantime, the differences between 'museums of nationalities' and 'museums of ethnology' were also subjected to examination. Wu Zelin noted that both types of museums are public specialty museums. The difference is that museums of ethnology are designed for research while museums of nationalities serve political propaganda, which are 'specialty museums related to ethnic minority groups' (Wu 1985). Wu suggested that the political function of museums of nationalities had been strengthened, but it required the involvement of more diverse disciplines.

In 1987, various experts, including anthropologists Wu Zelin, Yang Chengzhi, and Yang Kun, historian Bai Shouyi (白寿彝), archaeologist Su Bingqi (苏秉琦), and Ma Yin (马寅), the head of the preparatory group, attended a symposium on the preparatory work of the Chinese National Museum of Nationalities convened by the National Ethnic Affairs Commission. A consensus was reached that this museum should be a multidisciplinary complex institution, and the name 'Museum of Nationalities' was temporarily used.

In the 1980s, due to the severe loss of ethnic cultural relics in the pursuit of modernisation and the failure to gain approval for the construction of the museum, the focus remained entirely on collecting cultural relics and making applications for construction. For the third-generation practitioners of the museum of nationalities, the most important task was to master scientific methods for collecting and preserving national ethnic relics (Fei 1986, 1–2). In November 1985, the staff of the museum went to Hainan to collect cultural relics of the Li ethnic group (黎族). Fei Xiaotong carried out training programmes for them on collection and research in Guangzhou (Fei 1986, 1–2).

Explaining its importance and legitimacy is a crucial task for a museum. Yet the financial difficulties of China in the initial period of reform and opening up, and restrictions on 'erecting new government buildings' in the 1980s were both important reasons for delaying the establishment of the museum. In May 1995, the State Council officially approved the proposal for a Chinese National Museum of Nationalities. However, because the National Museum of China, also a national-level comprehensive museum, was still in the planning stage, it was felt the construction of the Chinese National Museum of Nationalities, had to be temporarily suspended 'until the content of the National Museum was finalized' to 'avoid repetition'. In the following years, faced with the coexistence of the museum under the CPN and the National Museum, the proposers of the Chinese National Museum of Nationalities had to argue repeatedly for what a museum showcasing ethnic groups of China should be, and how it could be used properly. The museum used the concept of 'the Chinese people of all ethnic groups uniting in diversity' proposed by Fei Xiaotong in 1988 (Fei 1988) to support its exhibition. By doing so, it made a distinction between its functions with those of the National Museum, the CPN museum, and all the museums of nationalities at different levels, combining its political responsibility with its pursuit of long-term historical continuity. The long-term failure to establish the Chinese National Museum of Nationalities led to a continuous loss of professionals, and the museum was gradually marginalised in academic anthropology.

The twenty-first century: anthropology of civilisations and museums of ethnology

Before receiving approval by the central government to provide a building, the Chinese National Museum of Nationalities had to continue its collection and protection of ethnic cultural relics, while revising its project proposal and clarifying its function. In this process, the debate over the concepts of museums of nationalities and museums of ethnology was brought into the twenty-first century.

In 2006, Shi Jianzhong (石建中), the founder of ethnographic museology in Minzu University of China (中央民族大学), defined museums of nationalities as museums showcasing ethnic minority groups in China or reflecting the histories and cultures of ethnic regions, which can be divided into four categories: (1) national-level comprehensive museums of nationalities; (2) local museums of nationalities; (3) museums of ethnology; and (4) specialty museums (Shi 2006, 45–6). This way of classification summarised the previous historical stage, considering a national-level museum of nationalities to be comprehensive and a museum of ethnology as being for academic purposes. In the same year, however, Su Donghai proposed the idea of 'museums with

Chinese characteristics showcasing ethnic groups' (Su 1986). He had argued in 1986 that there were two systems of museums in the world, namely the 'Eastern' and the 'Western'. At this point, he firmly believed that China's museums of nationalities should be built based on China's culture and status, with their own disciplinary system (Su 1986, 18).

In 2010, based on nearly 30 years of practice at the Chinese National Museum of Nationalities, scholars began to reconsider the name and academic background of the museum. Pan Shouyong noted:

> Museums of nationalities in the Chinese context are museums of anthropology (ethnology) in the Western context, and the absence of anthropology keeps museums of nationalities away from regulations and makes it difficult for these museums to develop academic influence and social status. (Pan 2011, 9)

Zheng Qian (郑茜), the deputy director of the Museum, stated in 2014 that the museum needed to find theoretical legitimacy, and believed that the name of the museum should be translated as the 'Chinese National Museum of Ethnology' (thereafter its official name in English) to express its desire to connect with anthropology in the international arena (Zheng 2014, 19). Tan Chee-Beng pointed out that 'ethnographic museums' are directly related to anthropology and that the Chinese National Museum of Ethnology, if established, would be the largest of its kind in the country (Tan 2016, 40).

The debate over the name of the museum in the twenty-first century was sparked off by the increasingly marginalised status of the museum, in specialising in history and archaeology, and from the difficulties experienced in collecting and displaying cultural relics. A more critical reason is that the museum had, as the theoretical basis for explaining its function, actively chosen the concept of 'the unity in diversity of the Chinese nation' developed by anthropologists.

In the second decade of the twenty-first century, the Chinese National Museum of Ethnology still does not have a physical building. For its legitimacy, the museum must first meet the requirements of shaping national unity. In this period, the influence of civilisation studies has begun to emerge. The anthropology of civilisations in the Chinese anthropological context follows the principle of 'prospering together' put forward in Fei Xiaotong's *The Pattern of Diversity in Unity of the Chinese Nation* (Fei 1988/2003), and the concept of 'supra-societal systems' introduced by anthropologist Wang Mingming (王铭铭) (Wang 2015). The development of the anthropology of civilisations in China provides a new method for explaining the formation of the Chinese nation. In the late 1990s, the guidelines for exhibition planning for the Chinese National Museum of Ethnology were 'to respond to the call of the International Council of Museums and collect today for tomorrow'.[9] Exhibitions were mainly to showcase the cultural heritage of ethnic groups.

> The process of the people of various ethnic groups and their ancestors living, working, reproducing, and creating colorful ethnic cultures since ancient times is shown by introducing the natural ecological environments inhabited by 56 ethnic groups to reveal the trajectory of the formation and development of the Chinese nation.[10]

Fei Xiaotong's concept of 'the pattern of diversity in unity of the Chinese nation', proposed in the early 1990s, had strong resonations within and beyond the fields of anthropology and ethnology. The Chinese National Museum of Ethnology revised and submitted a new project proposal in 2011.[11] The goal of exhibitions was changed to 'the history of the formation of the Chinese nation featuring unity in diversity supported by historical items'. The subsequent revised project proposal in 2020 proposed to 'present China featuring unity in diversity through physical evidence'.[12] In these new proposals, the entertainment and commercial aspects planned for the museum during the 1990s were removed, and the functions of protecting ethnic cultural relics and showcasing the rich Chinese culture proposed in the 1980s were retained. In addition, the function of expressing the long-term historical continuity and cultural commonality of the Chinese nation through exhibitions was strengthened. Through the use of anthropological analysis as its framework, the Chinese National Museum of Ethnology distinguished between its own functions and those of national museums, the Cultural Palace of Nationalities and other museums of nationalities by explaining the nature of unity in diversity of the Chinese nation through exhibitions.

To explain 'the pattern of diversity in unity of the Chinese nation' through museum exhibitions, the Chinese National Museum of Ethnology had first to introduce the process by which different ethnic groups and regions move toward unity under a grand historical framework, while analysing the integration of ethnic groups and their communication from a cultural perspective. However, there are many difficulties in integrating constantly changing and uncountable groups of people into one grand narrative of history. Most ethnic minorities do not have written languages or historical documents as the basis for their narratives. If 'the unity in diversity of the Chinese nation' needs to be displayed on a historical and archaeological basis, the whole process has to rely on historical documents in Chinese. Anthropology has long focused on the cultural interactions of ethnic groups as reflected in material culture, which can complement historical and archaeological materials with ethnographic materials and interpretative analysis, thereby achieving a complete narrative of the development of Chinese civilisation.

The Chinese National Museum of Ethnology has started to network with anthropologists all over the world by holding international conferences and workshops, and by making exhibitions based on ethnographic methods and responding directly to concerns in the field of anthropology. For example, in *Search of the Footprints of Reindeers: An Exhibition of Reindeer Culture in Northeast Asia in the Pan-Arctic Circle* held in 2015 in a university gallery in Yinchuan, Ningxia showed the impact of climate change and the northward shift of the tundra on the reindeer culture, with the focus on the connection and communication, diversity and consistency among 19 reindeer-related peoples in China as far as the Arctic Circle. *Tradition @Present: Timeless Style of Chinese Ethnic Attires* was curated in 2016 using a strategy of cutting the existing narrative threads created by CNP, and introducing the reflections on the relationship on tradition, the present and the future by various ethnic groups in China exhibited in their ethnic costumes. These exhibitions, which unravel culture from anthropological perspectives and are profoundly interpretive, go far beyond the propaganda of the 1950s and the simple introduction of ethnic knowledge in the 1980s. These exhibitions together show the efforts made by the Chinese National Museum of Ethnology to express the interaction between different ethnic groups, and even different civilisations, by re-applying an anthropological approach. These exhibitions have enabled the Chinese National Museum of Ethnology to receive widespread attention in the fields of anthropology, museums and fine art.

Conclusion

What should a museum which showcases the different ethnic groups within China be called? The century-long debate discussed in this chapter reveals the reflections among Chinese scholars on whether anthropology from the West can be used to interpret the population composition and cultural formation process featuring unity in diversity within China.

The debate stems from a particular historical context. China's civilisation that had existed for several thousand years, was disrupted after the late Qing Dynasty by foreign aggression, and 'China' as a modern state replaced 'Tianxia' (天下, meaning 'the world under heaven') with the move from dynastic rule to a modern nation with a clear state boundary and sovereignty. The cultural boundaries constructed by *Huayizhibian* (华夷之辩), the distinction between central China and non-Han ethnic groups around it, were changed accordingly, giving birth to a new sense of nationhood. Intellectuals at that time believed that it was imperative to impart knowledge about ethnic composition to the public and provide national education through museums. Therefore, they adopted anthropological theories to study and showcase ethnic groups within China.

Chinese anthropologists quickly reached a consensus through field surveys on both 'cultural diversity' and 'national unity'. However, they also found that Western theories on the relationship between ethnic groups and the state were insufficient to explain China's ancient civilisational system. As a result, they attempted to use museums to explain the consistency in the historical and cultural aspects of China as a multiethnic country. There is a persistent effort by Chinese anthropologists to develop a Chinese anthropological system. The specific name of the Chinese National Museum of Ethnology is a product of the integration of the Chinese anthropological system and the political function of the museum.

Despite its turbulent history, the development of anthropology in China has played a decisive role in clarifying the functions and exhibition planning of the Chinese National Museum of Ethnology. From 'the unity in diversity of the Chinese nation' proposed by Fei Xiaotong in the late-twentieth century to the development of the anthropology of civilisations in the twenty-first century, the path of a 'Chinese anthropology' has been reflected in the exhibition planning of the Chinese National Museum of Ethnology, which continuously expands and even redefines the scope of 'ethnological exhibitions' in China.[13]

Notes

1 Chinese characters are provided for the first mention of a name, where relevant.

2 国立中央博物院筹备处概况 ['Overview of the Preparatory Office of the National Central Museum'], edited by the Preparatory Office of the National Central Museum, 1942.

3 年国家民委召开关于中国民族博物馆筹备工作的专家座谈会会议纪要 ['Minutes of the symposium on the preparatory work of the Chinese National Museum of Ethnology convened by the National Ethnic Affairs Commission']. Materials collected in the Chinese National Museum of Ethnology.

4 中国人民共和国文化部对故宫民族文物调拨的意见 ['Opinions on the allocation of ethnic cultural relics from the Palace Museum'], 21 March 1959. The nationwide socio-historical survey also helped collect a large number of ethnic cultural relics.

5 民族文化宫落成和十年来民族工作展览 ['Inauguration of the Cultural Palace of Nationalities and exhibition of the work on ethnic affairs in the past 10 years']. Internal document of the Cultural Palace of Nationalities,1959.

6 全国民族工作展览(综合馆)资料汇编 ['National Exhibition of the Work on Ethnic Affairs (the comprehensive hall)'], in the Exhibition Hall of the Cultural Palace of Nationalities. Internal materials of the Cultural Palace of Nationalities, October 1980.

7 The Five Series of Books on Ethnic Issues includes *China's Minorities*, *A Brief History of China's Minorities*, *A Brief History of China's Minority Languages*, *An Overview of China's Minority Self-Governing Places*, and *A Series of Social and Historical Survey Materials on China's Minorities*. The series was edited under the auspices of the State Ethnic Affairs Commission of the People's Republic of China. The books were published separately by publishers in each province. The entire series consists of more than 300 volumes with nearly 50 million words.

8 全国少数民族文物工作会议纪要 [Minutes of the National Conference on Cultural Relics of Ethnic Minority Groups], 28 November 1984.

9 Chinese National Museum of Ethnology, ed. *Project Proposal for the Chinese National Museum of Ethnology, 2006*. Archived in the Chinese National Museum of Ethnology, p. 3.

10 Chinese National Museum of Ethnology, ed. *Project Proposal for the Chinese National Museum of Ethnology, 2011*. Archived in the Chinese National Museum of Ethnology, p. 1.

11 *Project Proposal for the Chinese National Museum of Ethnology, 2011*. Archived in the Chinese National Museum of Ethnology.

12 *Project Proposal for the Chinese National Museum of Ethnology, 2020*. Archived in the Chinese National Museum of Ethnology.

13 I am extremely grateful to Professors Graeme Were and Nick Stanley for the editing of this chapter, and Dr Zhang Lisheng for his help with the language. Thanks also to Professor Tan Chee-Beng for his invaluable advice with this chapter.

References

Cai, Yuanpei. 1921/1984. 'What is culture?'. In *Complete Works of Cai Yuanpei* (Vol. 4), edited by Gao Pingshu, 10–15. Beijing: Zhong Hua Book Company.

Chen, Guoqiang. 1980. '论民族学研究的对象 [On the objects of ethnological research]'. In 民族学研究第一辑——首届全国民族学学术讨论会论文集 [Ethnology Research Series I – Proceedings of the First National Symposium on Ethnology], edited by the Ethnological Association of China, 80–6. Beijing: Min Zu Press.

Chen, Kejin. 1997. '从民族博物馆名称说开去 [Starting from the name Museums of Nationalities]'. In 吴文藻纪念文集 [Collected Articles of Wu Wenzao], edited by Wang Qingwen, 106–11. Beijing: China Minzu University Press.

Chen, Mengjia. 1954. '中华民族文化的共同性 [The commonality of Chinese national culture]'. 文物参考资料 [Cultural Relics Reference] 9: 67–9.

Denton, Kirk A. 2013. *Exhibiting the Past: Historical memory and the politics and ideology of museums in the People's Republic of China.* Honolulu: University of Hawaii Press.

Duan, Yong. 2017. *Dangdai Zhongguo bowuguan* [Museums in Contemporary China]. Nanjing: Yilin Press.

Fan, Ke. 2008. '他者的再现与国家政治 [The reproduction of the Other and national politics]'. 开放时代 [Open Times] 6: 97–108.

Fan, Ke. 2016. 'Representation of ethnic minorities in socialist China'. *Ethnic and Racial Studies* 12: 1–17.

Fei, Xiaotong. 1956/1988. '关于建立中央民族博物馆的意见 [Opinions on the establishment of the Central Minzu Museum]'. In 费孝通民族研究文集 [Collection on Ethnic Studies of Fei Xiaotong]. Beijing: Ethnic Publishing House.

Fei, Xiaotong. 1986. '1985年11月1日在广州接见中国民族博物馆海南工作对全体队员时的讲话 [Speech delivered to all members of the Hainan Working Team of the Chinese National

Museum of Ethnology in Guangzhou on 1 November 1985]'. 民族文物工作通讯 [Ethnic Heritage Work Newsletter], edited by the Preparatory Group of the Chinese National Museum of Ethnology 2: 1–2. Materials collected in the Chinese National Museum of Ethnology.

Fei, Xiaotong. 1988/2003. 中华民族多元一体格局 [The Pattern of Diversity in Unity of the Chinese Nation]. Beijing: Ethnic Publishing House.

Hu, Qiaomu. 1984. '在中国博物馆学会迎春座谈会上的讲话 [Speech at the spring symposium of the Chinese Museum Society]'. *Museum*. April 1984: 5–7.

Jin, Tianming. 1980. '当前我国民族学的特点与任务 [The current characteristics and tasks of ethnology in China]'. In 民族学研究第一辑——首届全国民族学学术讨论会论文集 [Ethnology Research Series I – Proceedings of the First National Symposium on Ethnology], edited by the Ethnological Association of China, 67–79. Beijing: Min Zu Press.

Liang, Zhaotao. 1980. '我国应有自己的民族学 [Our country should have its own ethnology]'. In 民族学研究第一辑——首届全国民族学学术讨论会论文集 [Ethnology Research Series I – Proceedings of the First National Symposium on Ethnology], edited by the Ethnological Association of China, 12–22. Beijing: Min Zu Press.

Lin, Yaohua. 2014. '新中国的民族学研究与展望 [Ethnological research and prospects in New China]'. 民族研究 [Ethno-National Studies] 2: 48–55.

Liu, Bo'er. 2018. '费孝通中华民族多元一体格局理论的思想来源——民国时期民族学界"多元" "一体"理论设计研究 [Fei Xiaotong's ideological resources of the theory of "Chinese Nation with Multi-Ethnic Groups": Research on the theory design of "pluralism" and "unification" in the ethnological community in the Republic of China]'. 民族研究 [*Ethno-National Studies*] 3: 56–63.

Liu, Zhiqing. 1999. '民族文化宫博物馆的功能及其特征 [The function and characteristics of the museum under the Cultural Palace of Nationalities]'. In *Theory and Practice of Museums of Nationalities*, edited by the Cultural Palace of Nationalities, 19–26. Beijing: Ethnic Publishing House.

Lu, Tracey L. D. 2014. *Museums in China: Materialized power and objectified identities*. London: Routledge.

Ma, Xuming. 1999. '浅议民族文物的搜集 [A brief discussion on the collection of ethnic cultural relics]'. In 民族博物馆的实践与理论 [The Practice and Theory of Museums of Nationalities], edited by the Cultural Palace of Nationalities, 101–4. Beijing: Ethnic Publishing House.

Ma, Yin. 1987. '国家民委召开关于民博筹备工作的专家座谈会马寅发言稿 [Speech on the National Ethnic Affairs Commission holds a symposium of experts on the preparation of the museum of nationalities]', 21 February 1987. Archived by the Chinese National Museum of Ethnology.

Ma, Yin. 1988. '加强对民族文物的保护和研究工作——在全国少数民族文物工作会议上的讲话 [Strengthening the protection and research of ethnic cultural relics – Speech at the National Conference on Cultural Relics of Ethnic Minority Groups]'. 云南民族大学学报(哲学社会科学版 [*Journal of Yunnan Minzu University(Philosophy and Social Sciences Edition)*] 3: 12–16, 39.

Niu, Aihong. 2017. 'A study on the first ethnological exhibition in modern China: Models of Chinese and foreign nationalities held by Henan Museum of Ethnology in 1928'. 中国国家博物馆馆刊 [*Journal of National Museum of China*] 7: 100–11.

Pan, Shouyong. 2011. '民族博物馆是什么 [What is a museum of nationalities?]'. 中国民族报 (*Zhongguo Minzu Bao*), 29 October, 9.

Sa, Kongliao. 1959a. 在国庆展览工作会议上的讲话 ['Remarks on the National Day Exhibition at a Work Conference']. 31 March. Edited by Guo Yimei. Archived by Cultural Palace of Nationalities.

Sa, Kongliao. 1959b. 民族宫的问题 ['Speech on the Cultural Palace of Nationalities']. 14 January. Edited by Wuye. Archived by Cultural Palace of Nationalities.

Shi, Jianzhong. 2006. 民族博物馆学教程 [A Course in Museums of Nationalities]. Beijing: China Minzu University Press.

Sturtevant, William. C. 1969. 'Does anthropology need museums?'. *Proceedings of the Biological Society of Washington* 82: 619–49.

Su, Donghai. 1986. 'There are two systems of museums and museology in the world'. *Chinese Museum* 3: 26–7.

Su, Keqin. 1959. '关于少数民族社会历史调查的十年 [A decade of social and historical research on ethnic minorities]'. 民族研究 [*Ethno-National Studies*] 10: 42–7.

Tan, Chee-Beng. 2016. '博物馆、文化遗产与华侨华人研究 [Museums, cultural heritage, and studies on overseas Chinese]'. 西北民族研究 [*N. W. Journal of Ethnology*] 2: 39–48.

Tao, Yunkui. 1941/2017. '国立边疆博物馆计划 [Proposal for the National Borderland Humanities Museum]'. In *Ethnological Survey in the Period of the Southwest Associated University Period*, edited by Nankai University Museum, 206–10. Tianjin: Nankai University Press.

Varutti, Marzia. 2014. *Museums in China: The politics of representation after Mao*. Woodbridge: Boydell Press.

Wang, Mingming. 2015. *Supra-Societal Systems: Civilization and China*. Shanghai: Sanlian Bookstore Press.

Wang, Youqiao. 1936. 'Museums should focus on the department of nationalities and the scope of its research'. 河南博物馆馆刊 [*Henan Museum Journal*] 2: 1–5.

Wang, Zhaowu. 1982. '谈民族学博物馆 [On museums of ethnology]'. 民族学研究 [*Study of Ethnology*] 3: 283–7.

Wu, Jinghua. 1984. '全国少数民族文物工作会议开幕词 [Opening remarks of the National Conference on Cultural Relics of Ethnic Minority]', 27 October 1984. Archived by the Chinese National Museum of Ethnology.

Wu, Zelin. 1985. '论博物馆、民族博物馆与民族学博物馆 [On Museums, Museums of Nationalities, and Museums of Ethnology]'. 中南民族大学学报 [*Journal of South-Central University for Nationalities*] 3: 73–82.

Xu, Jian. 2016. *Great Foundations: The intellectual history of early Chinese museums*. Beijing: Science Press.

Yang, Kun. 1986. '谈谈民族学博物馆学 [About ethnographic museography]'. 中国博物馆 [*Chinese Museum*] 10: 28–33.

Yin, Lin. 2007. '民博工作回忆录 [Memoirs of the work of the museum]'. In 我们共同的期待：中国民族博物馆筹建23年纪念文集 [Our Common Expectations: A collection of articles commemorating 23 years of the Chinese Ethnic Museum], edited by the Chinese National Museum of Ethnology, 191–9. Materials collected in the Chinese National Museum of Ethnology.

Zeng, Zhaoyu. 1956/2009a. '中国西南部及南部少数民族文物展览会特刊序言 [Preface to the Special Issue of the Exhibition of Cultural Relics of Minorities in Southwestern and Southern China]'. In 曾昭燏文集·博物馆卷 [Zeng Zhaoyu's Collected Works – Museum Volume], edited by Nanjing Museum, 93–4. Beijing: Cultural Relics Press.

Zeng, Zhaoyu. 1956/2009b. '中央民族博物馆筹备处对于国内各兄弟民族文物的搜集范围 [The scope of collecting cultural relics from ethnic groups in China by the Preparatory Office for the Central Museum of Nationalities]'. In 曾昭燏文集·博物馆卷 [Zeng Zhaoyu's Collected Works – Museum Volume], edited by Nanjing Museum, 57–63. Beijing: Cultural Relics Press.

Zetterstrom-Sharp, Johanna. 2013. 'The future of ethnographic museums: The Pitt Rivers Museum, Oxford 19–21 July 2013'. *Anthropology Today* 29 (6): 27.

Zhao, Zhiyan. 2006. '民族文化宫筹建始末 [Preparations for the building of the Cultural Palace of Nationalities]'. 中国民族报 (*Zhongguo Minzu Bao*), February 2.

Zheng, Qian. 2014. '民族博物馆？民族学博物馆？——对一种独特的中国博物馆现象的理解与阐释 [Minzu Museum? Ethnographic Museum? Understanding and Interpreting a Unique Chinese Museum Phenomenon]'. 中国民族博物馆研究 [*Museum Anthropology Readers*] 2014 (1): 2–19.

Part III
Future directions

9

Indonesian youth practices in creating a media museum to preserve the sense of nationalism in a digital age

Endah Triastuti

Museums in Indonesia resonate strongly with the state's idea of nationalism (Anderson 1983; Billig 1995; Hitchcock 2005; Kusno 2012). As a multicultural country, the state promotes the artificial consciousness of a unitary and unifying identity (Kitley 2014) to protect the nation from a permanent threat from 'primordial cultures' (Anderson 1999b) that threaten social and political disintegration. In previous research (Triastuti and Rakhmani 2011) we showed that within the Indonesian context the state employs museums to discipline the sense of nationhood, embedding in the minds of its citizens the notion of the nation as a territorial space that dismisses parochial sentiments in favour of the wholeness of 'Indonesia'. We examined how Suharto's regime created the image of 'one-ness' or 'we-ness' fomented by the authoritarian power of the state.

However, Indonesian national everyday practices have changed rapidly since the end of Suharto's New Order regime. This shift significantly disrupts the dominant discourse that links nationalism to the museum (Bertrand 2004; Riddell 2002; Van Klinken 2002). There are at least two factors that challenge the dominance of the state. The first is the decline of centralised cultural policies post the New Order. This provides opportunities for provincial governments to re-think cultural policies (Jones 2012). The second is the growing infrastructure of new media in Indonesia that provides Indonesians with a contemporary space to rearticulate the meaning of democracy (Triastuti 2017).

Recognising the dynamic nature of museum studies (Fuchs 2019; Witcomb and Message 2013), this chapter notes the rapid attempts to appropriate technology into museum practices. The growth of internet penetration and the increasing popularity of mobile technology require museums to make the most of technology, especially in providing the public with more information (Besser 1997) but also in making people think about the relationship between the museum and visitors in a new and different way (Lang et al. 2006).

After the first wave of internet-based museums in the 1990s (Gaia et al. 2020), several museums began to digitally showcase physical artefacts and collections on websites (McTavish 2006). The next generations of digital museums provided visitor-oriented facilities including interactive maps (Boiano et al. 2012), multi-language support, augmented reality (Woolley et al. 2021) and multimedia exhibitions (McDaid et al. 2011).

A number of scholars agree that in this media-saturated era, museums need to compete, fighting for audience attention (Simon 2010). However, although new technologies have transformed the way museums perform their functions, there remain scholarly concerns that advances in technology will not significantly change visitors' behaviour, especially relating to the phenomenon of 'active dozing' (Kawashima 1998; Schweibenz 1998; Waterton 2010). Heiner Treinen (1993) introduces the notion of active dozing, whereby uninformed visitors explore the museum in a state of induced daydreaming, characterised by purposelessness and planless activity. According to Treinen, active dozing is encouraged by misguided museum curators whose tendency is to treat visitors as a mass media audience, rather than encouraging a participatory culture. Additionally, museum curators still primarily employ new media to create a one-way communication (Fletcher and Lee 2012) that controls the visitors' gaze (Casey 2003), so creating a disciplinary space to control knowledge, especially through curatorial authority (Bennett 1990; Chen 2013; Colwell 2015).

This chapter considers the notion that participatory culture is central to the vision of beyond-the-wall museums in a media-saturated world. I pay attention to the discursive power of the museum, which governs people's connection between the past and the future (Bennett 2017; Casey 2003). I argue that the discourse on museums would benefit from an overarching rethink of the notion of visitors' participation. This is important, especially within the context of Indonesia, where museums play a central role in promoting a sense of nationalism and in preserving Indonesia's authentic culture.

This chapter presents a digital ethnographic study of Indonesian youth's engagement with social media, which aligns with recent scholarly ideas of media use embedded in everyday culture (Couldry and McCarthy

2004, 119; Hjorth 2008, 73; Takahashi 2010, 30–1). It explores how Indonesian youth engage in everyday experience, imagination and ideas on nationalism through engagement with social media. I examine the intersection of current new media landscapes, youths' engagement with digital media in online spaces and the creation of the 'media museum' (Russo 2012; also as linked to 'production of space', Lefebvre 1991) through participatory culture as part of everyday nationalism (Knott 2015). In addition, I argue that youths' engagement with social media significantly subverts the previous operations of power structures. My aim is to answer the question 'what kind of things do youths do with their experience, imagination and ideas of the past to recontextualise the present in their engagement with media?' (Bird 2003; Takahashi 2010).

Challenging theoretical assumptions: nationalism, new media and the museum

A large number of studies of nationalism define nationhood by referring to shared territories (Hastings 1997), language (Croucher 2003), historical lineament (Smith 2009), culture (Heater 2016), authenticity (Fishman 1968) and sense of belonging (Anderson 1983). The various accounts of nationalism agree that there is no single definition but, rather, a complex and textured – yet flexible and dynamic – set of perspectives necessary to understand nationalism (Schlesinger 1987). For Lefebvre, nationhood is an abstract space, a normative space without transparency in the making (1991). According to him, the idea of 'nationhood', as for other abstract spaces, is highly manipulative in nature. It is governed by all kinds of authorities (Lefebvre 1991, 51), and thus tends to overthrow and to silence native knowledge and lived experience through the process of homogenisation (Butler 2012). Lefebvre's idea is in line with other scholars who project nationalism as imaginative (Anderson 1999a), contrived (Gat 2012) and socially engineered (Hobsbawm and Ranger 2012). Hobsbawm (2012), for example, insists that the notion of nationhood needs a careful exposition. He suggests that nationalism is structured by a mindless arbitrary consensus against the backdrop of the elites, who generate the artificial meaning and experience on national affiliations (see also Billig 1995).

Some scholarship argues that new media will shift museums into a democratising space (Barrett 2012; Pierroux et al. 2020). In this chapter, I argue that such scholarship ignores the glaring fact that the museum landscape is a curated stage. As a representation of space, the museum imposes a simulation of reality and eliminates different knowledge, silencing the expression of peripheral experiences (Lefebvre 1991). On the contrary, new media encourage users to find their own way to empower themselves and create a sense of agency (Triastuti 2013). New media provide an alternative space for a much greater diversity of voices and perspectives, including subversive and marginalised knowledge and expression. Furthermore, 'Web 2.0' recalls Abercrombie and Longhurst's (1998) idea of a diffused audience, and opens up the possibilities for users to be producers at the same time (see, for example, Bruns 2008; Hjorth 2008). It makes the internet a space with 'a range of new … architectures of participation' (Gane and Beer 2008; O'Reilly 2007). Relevant to this, Harold Jenkins (1992) separates spectatorship, or cultural window-shopping, from participation. In this context, he draws on Raymond Williams's notion of culture, which emphasises human agency within the context of everyday life. According to Jenkins (1992), participatory culture invites users to reappropriate available text, turn it into a raw material and then actively participate in creating and circulating the 'new product'. Christian Fuchs emphasises that participatory culture involves a range of activities including 'sharing, co-creation, remixing, reuse and adaption of content' (Fuchs 2021, 53).

I argue that social media platforms facilitate people's agency in rearticulating a sense of nationhood (Toyoshima et al. 2022). I question the dominant idea that the museum acts as a 'virtuous' institution to restore and to display artefacts in an attempt to commemorate history (Silverstone 2012) and to transmit knowledge (Black 2011). I maintain that the internet creates opportunities to disrupt the hegemonic power of the museum in narrating the history and knowledge of the nation by stimulating the emergence of representational space.

Additionally, this chapter questions the everyday taken-for-granted routines of nationalism, by exploring the subversive practices of Indonesian youths engaging in their own versions of nationhood through their engagement with digital media. Following Fuchs's argument on nationalism in the era of the internet (2019), I suggest that the everyday digital media practices reproduce indiscriminate contemporary modes of nationhood (Lekakis 2017) that reshape dominant knowledge and experiences of nationalism (Goode 2021; Jones and Merriman 2009). This study appraises the notion of everyday nationalism (Knott 2015) derived from people's lived experience, expressed in many ways (Hobsbawm, 2012; 2021; Knott 2015).

Contemporary discussions of museums consider forms of subversive practice provided by digital media. However, most studies remain focused on audience experience. In this chapter, I take up Russo's argument, which shifts the focus to an experiential model based on knowledge exchange and active cultural participation, and employ the theoretical assumptions of 'media as a practice' (Couldry 2004). According to De Certeau (1984), such practice is a response to marginalisation, which takes place in the local context in daily interactions. Thus, even though social practice is routine (Couldry and Hobart 2012), it will vary across time and place because practice constantly involves various interrelated factors. Practice reflects one's knowledge, practical understanding, rules and general acceptance in society (Loscher et al. 2019). Thus, practice can be observed in the form of collective action (Schatzki 2005) involving the media (Couldry and Hepp 2013; Couldry Hobart 2012).

Indonesian youth's sense of nationhood through a focus on their use of digital media

In this chapter, I argue that Indonesian youth's sense of nationalism can be observed from shared understandings and disagreements, implicit cultural understandings, informal knowledge, attitudes and dispositions embodied in their media practices (Côté-Boucher et al. 2014) at the level of everyday life (Bird 2003). Following Fuchs, this study sees youth as active but struggling meaning-makers of nationalism, who ascribe meaning through a series of practices involving consumption, production and appropriation to media content to articulate their own meaning from texts, reflecting their own history. This new vision of users links closely to Doueihi's notion of anthology (2011, 31) and to Russo's idea, already mentioned, of the 'media museum' (2012, 11), which points to personalised authorship, including of the meaning of nationalism. It proposes views of authorship and intentionality that not only identify the changing roles between readers and authors, but also the possibility of a personalised and autonomous model of authorship (Doueihi 2011, 31–4).

In researching these areas I used digital ethnography (Boellstorff 2008), designed to access data which explore Indonesian youth's sense of nationhood through a focus on engagement with digital media and their everyday use of it. Data were approached using practice theory (Couldry and Hobart 2012) and the 'media museum' (Russo 2012). Data collections explored the implication of the digital world in the everyday lives of Indonesian youth, strongly related to how their everyday use of digital media – their practices – are intertwined within experience, knowledge and practices of 'being national' (Pink et al. 2016). Data collection involved online observation, focusing on Indonesian youth's expression of nationalism, with a focus on mediated practices that related to this. From this came a discussion of the hierarchy of their engagement with digital media within the various contexts of the wider media-saturated world. Their digital media practices were strongly affected by a range of interrelated factors relating to nationalism, namely experience, knowledge and imagination.

I undertook in-depth interviews with 10 Indonesian youths who visited museums, three museum curators, one historian and two museum public relations officers. I sent interview requests by direct message to Indonesian youths whose digital media activities on Instagram I had observed. Of these, 10 responded to the request and agreed to be interviewed. Museum curators, historians and museum public relations officers were approached using snowball techniques. Several interviews were delivered online because the participants did not live in Jakarta, where I was based. These interviews inform and elucidate what follows.

Why do Indonesians visit museums? The governing power of the discourse on nationalism

In 1994, four years before the fall of the New Order regime in Indonesia, Christina F. Kreps completed a PhD dissertation, *On Becoming 'Museum-Minded': A Study of Museum Development and the Politics Of Culture in Indonesia* which examined the main character of Indonesian museums as being governed by the logic of a dominant top-down approach (Kreps 1994). Although the state also built provincial museums, the majority of local museums retain the narrative of nationalism imposed by the ruling regime. The expression of a sense of nationalism is strongly embodied in the narrative and arrangement of displays (McGregor 2003; Putri 2019), so providing an authentic nationalism through history, collections and dioramas. This strict curatorial process helped the state to guarantee a consistent approach. The state additionally required schools to undertake obligatory museum visits (Taum 2015; Van Klinken 2005).

However, the results of my research show that the guided democracy imposed by the state creates a sense of alienation among visitors which militates against the authentic knowledge and history provided by museums.

As these two youth interviewees explained:

> I visited a museum when I was in high school. I went there during the school's excursion. It was a compulsory programme, so I did not have any other option but to join. I was ecstatic not because we visited a museum, but because we had a trip. I did not have any knowledge about what I saw at the museum. I did not know it was a collection of ancient artefacts that tell stories about Indonesian history, yet we were asked to submit a critical essay about what the museum means to us and what we feel about our history. (University student, age 23, August 2022)

> I did not know that Museum Lubang Buaya is framed by the tragedy of G30S PKI [Gerakan 30 September Partai Komunis Indonesia, the 30 September Movement]. I did not experience sense of the sinister as I know now [was] experienced by many visitors. Then in my critical essay, that I was obliged to write, I looked at [information] from another resource, but I did not understand and feel anything. It is just an obligation. (Youth, age 20, May 2022)

My interviewees' stories concur with Kreps' findings. Museum visits have not instilled a sense of nationalism, as was always intended by the state. Cultural activities like museum visits that were so thorough under the New Order are now losing any meaning. After the fall of Suharto's regime in 1998, democracy and nationalism become the subject of multiple interpretations, and museum visits exhibit this shift. The Reform Era under B. J. Habibie withdrew various indoctrination programmes from the New Order regime, including the state doctrine of Pancasila (BP7). BP7 was a powerful ideological apparatus under the New Order that orchestrated the state's indoctrination programmes.

Have Indonesia's museums lost their sense of purpose?

As one interviewee put it:

> Museums now become a storage or a repository of artefacts. There is a huge pile of unmanaged artefacts. I think the state does not have any ideas who to assign in managing the collections. (University student, age 21, January 2022)

Furthermore, museums are now concerned about the fall in their visitor numbers. The abandonment of centralised cultural policy post the New Order has reduced the number of people visiting museums. Mandatory visits to several museums such as Museum Pengkhianatan PKI (the Museum of the Indonesian Communist Party's Treachery) and the Satriamandala Museum – are no longer part of a compulsory programme.

In the words of interviewed museum personnel:

> As a text, the museum's collections and displays are not for everyone. The role of curator in storytelling does not guarantee that museum visitors will understand if they do not think that it somehow relates to what they understand and experience. (Museum consultant, age 46, November 2022)

> Yes, we still have visits from schools. But I don't think school children have a proper prior knowledge about Indonesia's heroism and history. They just run around, show no interest in knowledge about our collections. (Museum public relations officer, age 43, April 2022)

> The decentralisation policy has also led the state to hand over the management to the regional authorities. This has also led to a fall in visitor numbers. At the same time, the state has stopped prioritising museums in national cultural policy.
>
> After Suharto's reign ended, the financial support for developing Indonesian museums has followed the logic of government decentralisation. Today's museums are funded by various stakeholders. Each stakeholder has a different motive in developing museums. Some of them do not follow the formal arrangement for developing museum policy which is too standardised to follow. (Museum curator and public historian, age 45, November 2022)

Another consequence is the decline in choice of museums to visit. The four most-visited museums in Jakarta in 2021 were: (1) the Jakarta History and Inscription Museum, with 51,952 visits, which had decreased by 64.4 per cent from 2010 when there were as many as 145,771 visits; (2) Onrust Island Archaeological Park, with 23,135 visits in 2021, down 17.9 per cent from the previous year which had reached 28,165 visits; (3) Wayang Museum, with 20,632 visits in 2021, down 57.4 per cent from the previous year's 48,456 visits; and (4) the Museum of Fine Arts and Ceramics, Maritime, Textiles, and the Joang '45 Museum.

My interviewees gave their personal preferences and reasons for visiting particular museums:

> I love to visit museums that are entertaining. The Jakarta History Museum has a wide-open space for doing various things despite only [being able] to see Indonesia's heritage. I think visiting a museum does not always have to be with intention to learn something. That would be my last intention. I am looking for leisure. (Student, age 23, November 2022)

> I don't feel like I am visiting a museum. To me, visiting it [the Onrust Island Archaeological Park] is more like experiencing ecotourism. Yes, we were explained about the Onrust Island's history. But we also learned about the biodiversity in the island. (Youth, age 19, April 2022)

> I was on a museum date [giggles ...]. It was a date with my girlfriend. We strolled and got bored, then we searched for somewhere where we could eat. (Student, age 20, August 2022)

Nowadays, Indonesians visit museums to experience leisure and entertainment, regardless of the sense of, and the knowledge about, nationalism offered by museums. Visitors are also free to choose which museums to visit. Indonesians are likely to visit museums where they can entertain family members. Museums are shifting from a focus on education alone towards an accommodation to leisure markets.

Indonesia's museums in a digital era: creating the public's encounter with heritage

As a result of the changes in the role of the museum in Indonesia the government decided to join an international network of museum organisations, the International Council of Museums (ICOM). ICOM has developed a reference tool that sets standards of excellence to which all members of the organisation must adhere. By applying these reference tools, every museum creates standardised arrangements and curatorial systems so that its display and content are universally understood, valued and protected. However, by joining ICOM, the core aim to develop museums within a national context is weakened.

A museum curator commented:

> There are many museums in Indonesia that are significantly keen to develop new programmes. During the setup of a new museum, they create a standardised system that includes curatorial processes in choosing collections and creating narratives that visitors can understand and like. As soon as the curatorial decision process has finished, museum employees are busy ensuring that they use standardised museum arrangements in terms of the percentage of open space, layout, affordability, and other technical arrangements. However, they forget to create future innovations, which are also important to invite visitors and give them a great experience. (Museum curator, age 44, May 2023)

A number of national museums have attempted to establish digital heritage, using social media, even prior to the COVID-19 pandemic. The growing trend to initiate digital heritage arises from the younger generations of museum curators who understand the importance of using social media as a space to invite engagement. It is also evident that the senior management within Indonesia's museums lacks knowledge and interest in employing social media to promote visitors' engagement with national heritage. As one museum employee described it:

> Yes, we do have an account on Instagram and Facebook. But they are unofficial. Personally, I created the account – on the behalf of the museum where I am working. The content production is intermittent. I create content when I have spare time. The recruitment system does not allow us to hire a content creator who specifically organises content production. (Museum public relations employee, age 43, April 2022)

There are a number of museum accounts on various social media platforms, such as Facebook, Instagram and Twitter (now X). However, online visitor engagement is very low. Although some museums constantly post content on the platforms, they still lack comments, views, likes and reposts. Some museums display a lack of confidence in creating spatial and social connections between audience and museum. Satria Mandala Museum, for example, has five unofficial accounts on Instagram, namely @museumsatriamandala, @museum.satria.mandala, @museum_satriamandala_postmo, @kesatriamandala and @pusjarahtni.idn. The museum has two unofficial accounts on

Facebook and one unofficial account on Twitter/X (@kesatriaman-dala). None of Satria Mandala's accounts on social media platforms are verified. However, although @museumsatriamandala on Instagram has 1,250 followers, it has very low engagement, only having made 27 posts. Additionally, the majority of posts do not invite online visitors' participation.

Elisa Giaccardi makes the bold statement 'heritage meanings and values are not attached to artefacts, buildings or sites' (2012, 2). Yet, historical and national heritage, including collections, buildings, sites and artefacts, feeds visitors with imagination of the past and with visions of the future (Setten 2005). In that sense, the museum as heritage, resembles a space for visitors' agency, especially in creating meaningful experiences. That is to say, in order to have meaningful imagination and vision within the museum, visitors are engaging in knowledge exchange and active cultural participation, which is highly contextualised.

The Satria Mandala museum creates a division between the museum and museum visitors by posing management as in-group that carries out campaigns. Although the museum employs advanced technology, it nurtures an authoritative one-way communication, as in its social media post, 'yuk, jadi Generali bansa yg ideal' ('let's be the ideal generation'). This post articulates a utopian narration that is peculiar and distances visitors from both the past and the future. It alienates them from meaningful participation with their heritage, and prevents them from mobilising themselves throughout cross-media interactions (Giaccardi and Palen 2008).

However, virtual heritage remains a challenging practice within Indonesia's museology. The museum management insists on maintaining its authoritative power, making the excuse that:

> ... our digital culture is very fragile. In this crazy era of social media, we rely on young people from the millennial generation and involve [them] as our digital team. But not all museums have such a vision. Thus, not all museums have a creative content team. (Museum curator and public historian, age 45, March 2023)

Within the Indonesian context, the urge to establish online museums emerged during the COVID-19 pandemic. Various forms of social isolation left everyone with no option but to remain at home. A most important aspect of the museum was eliminated – namely, visitors.

> During the pandemic, all museums were physically closed. As a result, museum staff started to initiate digital space and connections. Eventually, staff who would previously have physically worked in the museum were diverted to online activities. All museum people were outsourced, to establish virtual heritage. It includes creating content, managing social media accounts, and others. Thus, even though we were closed, yet digitally, we are open – even, now, on a 24-hour basis, because at any time anyone can access our social media account. (Museum curator and public historian, age 45, May 2023)

However, data from my research reveals that the museum remains an important dominant institution in nurturing dominant knowledge and memory of nationalism, history and a sense of patriotism. During their museum visits, Indonesian youths shared their experience on social media and thus showed the 'nurtured civilisation' (McGregor 2003) by visiting a museum to confirm knowledge and connections of the past. There are youths who reproduce the dominant knowledge and memory of nationalism through digital captions, hashtags and comments. These practices, to a great extent, nourish and reproduce the dominant narration on nationalism previously imposed by the New Order, creating online an abstract space for 'unity in diversity' (Pancasila) and patriotism.

While visitors may wish to recreate the memory of the horrific episodes of the 30 September Movement as part of Museum Pancasila Sakti's virtual heritage on its Instagram account, @monumenpancasilasakti invites no participatory activities, instead guiding the visitors' acceptance of the explanation offered and seeking to create a sense of nationalism. As an example of virtual heritage, @monumenpancasilasakti demonstrates a visual focus that affected visitors' everyday nationalism. The content is highly manipulative in nature and offers mainly abstract concepts of nationalism, which creates a sense of alienation among today's youth, who were born after 1965. Visitors' comments are also homogenous. This is an example of the massive mobilisation of the artificial meaning and experience of national identity.

The 'media museum' and participatory culture

Scholars suggest that advanced technology changes the nature of museums, especially with respect to knowledge production and visitor experience. There are at least two different views about the role of new media in museology. The first emphasises how the new media may bring a richer and more meaningful experience for museum visitors. This perspective has emerged against the backdrop of persistent critiques that museums enhance the one-way communication that encourages visitors to practice 'active dozing' – that is, visiting a museum without plans and knowledge.

Some scholars suggest that museum programmes are hypocritical. On one hand museums apply various strategies to increase the number of visits, such as developing the education room (Del Chiappa et al. 2013), creating online exhibitions (Kamariotou et al. 2021), providing data visualisation (Lanir et al. 2013), renovating display facilities (Kawashima 1998) and providing facilities for disabled people (Mesquita and Carneiro 2016). But on the other hand, museums still employ the interaction format of mass media communication in which 'museum visitors' are conceived of as a large number of anonymous individuals (Porsché 2022).

As one interviewee stated:

> Every museum should have sufficient knowledge about their visitors. Museums should be able to profile visitors so that all shared information in the display can be understood by the target audience. Right? I think they should. Museums should have an awareness that the material they have can only be understood by a certain age level. So, visitors are seen as a qualitative construct. Most museums perceive and treat visitors as numbers, using quantitative perspective. Often, I see many elementary school children in groups, running here and there in the museum because they don't understand what they see or what the museum provides them. (Museum curator, age 48, March 2023)

However, there has recently been a development from seeing the museum as a temple to that of a forum (see Cameron 2004). Despite relying on the curatorial elite, these museums invite external stakeholders to design various strategies to expand visitors' engagement with the museum. For example, the National Museum of Indonesia employs young creative workers, not only to design storylines but also to hire a third party to fulfil the museum's goal:

> The National Museum of Indonesia hired us. We created a programme 'Weekend at the Museum'. In that programme we revived a collection – whether it is an artefact, or a group of artefacts, we package it in the form of live theatre. We collaborate with Teater Koma, who perform the play. (Museum curator, age 44, January 2023)

When the National Museum of Indonesia allows external stakeholders to ascribe new meaning to their collections, it allows layers of meaning to be unpacked in a single artefact. Moreover, this new museum practice produces alternative meaning for an artefact:

> First, we will refer to the National Museum of Indonesia's annual theme. Once, the museum picked multiculturalism as its annual theme. Then, we'll look for the current state of affairs that fit that theme. We picked the national presidential election. At that time, it seemed that multiculturalism in Indonesia was in danger of being dropped. The next step is to look for collections that are in line with the theme. At that time, we decided the story of the Sultan of Banten's Crown fits with the theme. Banten Sultanate was one of the important international trading centres. It allowed cultural, ethnic, and religious diversity. Foreigners and local people assimilated, integrated and created social interaction. Overall, we considered [that], as an artefact, the Sultan of Banten's crown fitted with the context and theme. Following up, we delivered our research to find the ingredients to develop a story for the theatre to play. (Museum curator, age 46, March 2023)

Rumah Dongeng and Teater Koma involve participatory practice when deploying their contextualised knowledge and experience. They bring a voice from the margin into the dominant institution. Thus despite their subversive nature, inviting multiple external stakeholders modifies 'the between the walls place such as museum' into a space, 'which the subject may take up a position and speak of the objects with which he deals in his discourse' (Michael Foucault in Lefebvre 1991, 4). Although the museum is becoming more performative in nature, it remains a stronghold against visitors. Findings in this study are consistent with Valery Casey's argument, that museum practices in the digital era paints the image of a walled garden (2003, 16). Museums' various programmes, agendas and technological features can be only experienced within the museum's natural environment, which is highly managed. Despite the enjoyable performance that visitors experience, this kind of arrangement turns visitors into consumers who remain tied to 'authentic' (as in authoritative, undisputed) knowledge and nationalism.

Outside Casey's 'walled garden' there exists a scenery of more generative ecosystems. A number of youths on social media devote their time to reframing an understanding and experience of national history. They take up their personal preferences to generate value through their imagination, mental landscape, memory, sense, identities and experience. They convert their social media accounts into a form of historical preservation, one that is subversive yet genuine. It is subversive because these youths create new meaning and significance for the dominant historical story, transforming it into a meaningful heritage that connects them to both the past and the future.

A TikTok account, @elsa.novias is dedicated to displaying a collection of narratives about Chinese Indonesians' heritage. There is no museum in Indonesia that preserves and performs Chinese Indonesians' culture as a part of Indonesian history. Chinese Indonesians have always been a small minority. Under the New Order, Suharto created a discriminatory policy against those of Chinese descent in Indonesia to keep them second-class citizens. There are several political and social events where Chinese Indonesians suffered from harsh violence. Taking into account the profound discrimination, negative stereotype and marginalisation against the Chinese Indonesian community, @elsa.novias turns her TikTok account into a conceived space and, at the same time, a lived space. As a content creator, she takes advantage of TikTok's features to create awareness of the dominant constructed code and symbols revolving around Chinese Indonesians. TikTok's format, based on presenting sound and image, vividly presents Chinese Indonesians as inhabitants of a specific stereotyped space. By posting 'Kenapa orang Tionghoa mayoritas berdagang' [Why do the majority of Chinese Indonesians people trade], using the #serunyasejarah [history is exciting] tag, she brings the negative stereotype of Chinese Indonesians into the discussion.

The architecture of social media, like the hashtag, promotes this subversive meaning. She creates a story that turns the post into a lived space, by making sense of her memory as a member of the Chinese Indonesian community. Additionally, she develops a sense of identity through shared interaction using comments and reposts. This particular TikTok post received 7,141 likes, 167 comments, 1,047 bookmarks, 428 reposts and 117, 700 views.

The TikTok post portrays a form of participatory culture that emphasises the values of diversity and democracy rooted in interpretation, production, curation and circulation of meaning through active interactions; not only users' interaction with media but also their interaction with each other. TikTok is in line with the second view about the role of new media, that advances in technology can reduce dominant ideology of the museum and move heritage into a space produced by visitors – a space that derives from visitors' own experience, understanding and knowledge. This perspective is rooted in the scholarly argument that the human is an active meaning-maker (Schroeder 2007).

It is also important to note that @elsa.novias indicates that the internet enables a contest between different ideas about knowledge and memory in nationalism, history and a sense of patriotism. There are personal accounts on various social media platforms where users share creative 'DIY' posts to articulate their individual understanding of what national heritage can mean. Owners of such accounts are not museum staff, but nevertheless they interact with heritage objects and concerns. This leads to a transformation of display spaces, fields of practice and use of media – where the museum as a site of cultural heritage is appropriated into an experiential model based on knowledge exchange and active cultural participation (Russo 2012). However, I disagree with Neil Silberman and Margaret Purser's idea that the internet is a safe space for subversive meaning and storytelling, altering dominant knowledge and experience (Silberman and Purser 2012). On the contrary, within the Indonesian context, the internet has gradually enhanced the ideological legitimacy of the authoritarian regime.

Sejarah Seru (@SejarahSeru) is a site of virtual heritage that employs digital platforms to create a media museum. It owns one account on TikTok and one account on YouTube. @SejarahSeru has been created to develop a more enjoyable experience yet remaining focused on conveying an 'authentic' (authoritative) history. One of those involved explained:

> We work together to provide and to share authentic storytelling on history. Yet we also aim to provide the other face of history for users. You know what I mean, right? As we believe, our history is produced by the ruling elite. We are very serious. Yes, we do not interview people. But we use trusted, authoritative books as our source. One of us is also a historian. He will make sure that the information we will share is already accurate. (Museum consultant, age 46, January 2023)

Despite the belief that a media museum should be a representational space for engagement, participation and co-creation, it is evident that social media platforms remain linked to the ruling elite. Beyond the claim of 'timeless and spaceless', digital platforms have, surprisingly, a potentially detrimental impact on public interests, and can hinder innovation and damage the notion of participatory culture. A museum consultant described their experience:

> Last year, we posted our version of Surat Perintah Sebelas Maret[1] on YouTube. At that time, we believed that the internet provided us with democratic space. Our version of Surat Perintah Sebelas Maret is different to the official version. We understand that this is quite a sensitive content. Thus, we decided to create a storyline for an animation movie. A day later, we were shocked to find that the platform has taken down our content. We did try for the second time, and it was the same. (Museum consultant, age 46, January 2023)

Some digital platforms have now become mainstream media. My enquiry agrees with the findings of previous studies, that media conglomerates have become powerful forces in Indonesia (Tapsell 2012). In effect, the authoritarian regime has groomed the ideological state apparatus well, so that not only has the legacy of the New Order endured, but all the national ecosystems in the enterprise work together perfectly across time, nurturing this legacy.

However, there are major chinks in this authoritarian armour. As we have previously argued (Triastuti and Rakhmani 2011) Taman Mini Indonesia Indah (TMII) or the 'Beautiful Indonesia Miniature Park' is a physical artefact of grandeur, considered as a monument within the national culture of Indonesia. Built under Suharto's regime, TMII embodies this ideological past. In TMII, the uniqueness of each ethnicity in Indonesia is highly accentuated and framed within the museum complex. It symbolises the desire of the authoritarian regime (through the hegemony of nationhood) to control Indonesian local communities by providing a space that represents mental and physical boundaries. The Indonesian government maintained the locality of each ethnic group, allowing ethnic sentiments and nationalism to coexist. This visible ethnic 'territorial marking' is reinforced by institutional media, particularly through television, as part of a larger nation-building scheme. The introduction of the internet, on the other hand, has blurred these territorial markings. Data shows that in the rapid growth of local blogger communities – permitted by the increasing availability of the internet – they have instead taken these territorial 'Indonesian' markers as their online identities. These re-employed markers, which were first created by the authoritarian government, now merge together the notion of 'physical' and 'mediated' space. These blogger communities, not unlike TMII, are shaped and bordered by physical territorial names in cyberspace. Reading the list of names of local blogger communities is like reading the list of traditional houses built inside TMII. Blogging practices have allowed members to form a resistance, by information sharing and questioning other methods to 'hegemonise' the mosaic of ethnic groups that forms the Indonesian blogger community. Local culture operates to ensure community and/or society participation and furthermore the sustainability of networking. From this point of view, blogging practices cannot be treated merely as such. They are in fact pervasive cultural artefacts that form a blogging culture.

Conclusion

In media discourse, Indonesian youth apply contextual tactics to reclaim their sense of agency by experiencing struggle through their participation in online heritage, as embodied in various practices. Reframing museum practice, Indonesian youth modify the internet as a space into a heritage, where they re-narrate nationalism and Indonesia's history, employing their imagination and marginalised knowledge. Today, in an era of complex media, Indonesian youths create media rituals to describe society's engagements with media. In media museum practice, a media-saturated world 'relocates' the audience from 'the sitting room' to 'a variety of social contexts'. So, media power cannot be considered as direct and massive; instead it is dispersed in the 'whole spectrum of talk, action and thought that draws on media', and is varied, and depends on people's engagement with media (Couldry 2004, 196).

Heritage practices are never simple, limited or fixed. Even though previously museums were the most powerful institutions in their narrating of history and nationalism, this does not necessarily mean that visitors' experiences with history and nationalism have been, or are, simply as consumers who are powerless or passive. In their engagement with social media, Indonesian youth subvert and 'poach' history and nationalism by appropriating social media and reconstructing it for their own social, economic, political and cultural contexts and benefit. However, at the same time, Indonesian youth's engagement with social media has challenged the dominant power elite.

Note

1 A document signed by the Indonesian President Sukarno on 11 March 1966, giving army commander Lt. Gen. Suharto authority to take whatever measures he 'deemed necessary' to restore order to the chaotic situation during the Indonesian mass killings of 1965–6 (Source: Wikipedia).

References

Abercrombie, N. and B. Longhurst. 1998. *Audiences: A sociological theory of performance and imagination*. London: Sage.

Anderson, B. R. G. 1983. *Imagined Communities: Reflections on the origins and spread of nationalism*. New York: Verso.

Anderson, B. R. G. 1999a. 'The goodness of nations'. In *Nation and Religion: Perspectives on Europe and Asia*, edited by Peter van der Veer and Hartmut Lehmann, 197–204. Princeton: Princeton University Press.

Anderson, B. R. G. 1999b. 'Indonesian nationalism today and in the future'. *Indonesia* 1: 1–11.

Barrett, J. 2012. *Museums and the Public Sphere*. Hoboken, NJ: John Wiley & Sons.

Bennett, T. 1990. 'The political rationality of the museum'. *Continuum* 3 (1): 35–55.

Bennett, T. 2017. *Museums, Power, Knowledge: Selected essays*. London: Routledge.

Bertrand, J. 2004. 'Democratization and religious and nationalist conflict in post-Suharto Indonesia'. In *Democratization and Identity: Regimes and ethnicity in East and Southeast Asia*, edited by S. J. Henders, 177–200. Lanham, MD: Lexington.

Besser, H. 1997. 'The transformation of the museum and the way it's perceived'. In *The* Wired *Museum: Emerging technology and changing paradigms*, edited by Katherine Jones-Garmil, 153–70. Arlington, VA: American Alliance of Museums.

Billig, M. 1995. *Banal Nationalism*. London: Sage.

Bird, S. E. 2003. *The Audience in Everyday Life: Living in a media world*. London: Routledge.

Black, G. 2011. 'Museums, memory and history'. *Cultural and Social History* 8 (3): 415–27.

Boellstorff, T. 2008. *Coming of Age in Second Life: An anthropologist explores the virtually human*. Princeton: Princeton University Press.

Boiano, S., J. P. Bowen and G. Gaia. 2012. 'Usability, design and content issues of mobile apps for cultural heritage promotion: The Malta culture guide experience'. Accessed 10 September 2024. https://arxiv.org/abs/1207.3422.

Bruns, A. 2008. *Blogs, Wikipedia, Second Life, and Beyond: From production to produsage*. New York: Peter Lang.

Butler, C. 2012. *Henri Lefebvre: Spatial politics, everyday life and the right to the city*. London: Routledge-Cavendish.

Cameron, D. 2004. 'The museum, a temple or the forum'. In *Reinventing the Museum: Historical and contemporary perspectives on the paradigm shift*, edited by Gail Anderson, 61–73. Walnut Creek, CA: AltaMira.

Casey, V. 2003. 'The museum effect: Gazing from object to performance in the contemporary cultural-history museum'. *Archives & Museum Informatics* 2: 1–21.

Chen, K. T. N. 2013. 'The disciplinary power of museums'. *International Journal of Social Science and Humanity* 3 (4): 407–10. https://doi.org/10.7763/IJSSH.2013.V3.27.

Colwell, C. 2015. 'Curating secrets: Repatriation, knowledge flows, and museum power structures'. *Current Anthropology* 56 (S12): S263–75.

Côté-Boucher, K., F. Infantino and M. B. Salter. 2014. 'Border security as practice: An agenda for research'. *Security Dialogue* 45 (3): 195–208.

Couldry, N. 2004. 'Theorising media as practice'. *Social Semiotics* 14 (2): 115–32. https://doi.org/10.1080/1035033042000238295.

Couldry, N. and A. Hepp. 2013. 'Conceptualizing mediatization: Contexts, traditions, arguments'. *Communication Theory* 23: 191–202.

Couldry, N. and M. Hobart. 2012. 'Media as practice'. In *Media, Society, World: Social theory and digital media practice*, 33–58. Cambridge: Polity Press.

Couldry, N. and A. McCarthy. 2004. *MediaSpace: Place, scale, and culture in a media age*. London: Routledge.

Croucher, S. L. 2003. 'Perpetual imagining: Nationhood in a global era'. *International Studies Review* 5 (1): 1–24.

De Certeau, M. 1984. *The Practice of Everyday Life*. Berkeley: University of California Press.

Del Chiappa, G., M. G. Ladu, M. Meleddu and M. Pulina. 2013. 'Investigating the degree of visitors' satisfaction at a museum'. *Anatolia* 24 (1): 52–62. https://doi.org/10.1080/13032917.2012.762317.

Doueihi, M. 2011. *Digital Cultures*. Cambridge, MA: Harvard University Press.

Fishman, J. A. 1968. 'Nationality-nationalism and nation-nationism'. In *Language Problems of Developing Nations*, edited by J. Fishman, C. Ferguson and J. Das Gupta, 39–51. New York: Wiley.

Fletcher, A. and M. J. Lee. 2012. 'Current social media uses and evaluations in American museums'. *Museum Management and Curatorship* 27 (5): 505–21.
Fuchs, C. 2019. *Nationalism on the Internet: Critical theory and ideology in the age of social media and fake news*. London: Routledge.
Fuchs, C. 2021. *Social Media: A critical introduction*. London: Sage.
Gaia, G., S. Boiano, J. P. Bowen and A. Borda. 2020. 'Museum websites of the First Wave: The rise of the virtual museum'. In *Proceedings of EVA London 2020*, 24–31. BCS Learning & Development. https://doi.org/10.14236/ewic/EVA2020.4.
Gane, N. and G. Beer. 2008. *New Media* [electronic resource]. New York: Berg.
Gat, A. 2012. *Nations: The long history and deep roots of political ethnicity and nationalism*. Cambridge: Cambridge University Press.
Giaccardi, E. 2012. 'Introduction: Reframing heritage in a participatory culture'. In *Heritage and Social Media*, edited by Elisa Giaccardi, 1–10. London: Routledge.
Giaccardi, E. and L. Palen. 2008. 'The social production of heritage through cross-media interaction: Making place for place-making'. *International Journal of Heritage Studies* 14 (3): 281–97. https://doi.org/10.1080/13527250801953827.
Goode, J. P. 2021. 'Artificial intelligence and the future of nationalism'. *Nations and Nationalism* 27 (2): 363–76.
Hastings, A. 1997. *The Construction of Nationhood: Ethnicity, religion and nationalism*. Cambridge: Cambridge University Press.
Heater, D. 2016. *The Theory of Nationhood: A platonic symposium*. New York: Springer.
Hitchcock, M. 2005. '"We will know our nation better": Taman Mini and nation building in Indonesia'. *Civilisations* 52 (2): 45–56.
Hjorth, L. 2008. *Mobile Phone Culture in the Asia Pacific: The art of being mobile*. London: Routledge.
Hobsbawm, E. J. 2012. *Nations and Nationalism Since 1780: Programme, myth, reality*. Cambridge: Cambridge University Press.
Hobsbawm, E. J. 2021. *On Nationalism*. New York: Hachette.
Hobsbawm, E. and T. Ranger. 2012. *The Invention of Tradition*. Cambridge: Cambridge University Press.
Jenkins, H. 1992. *Textual Poachers: Television fans and participatory culture*. London: Routledge.
Jones, R. and P. Merriman. 2009. 'Hot, banal and everyday nationalism: Bilingual road signs in Wales'. *Political Geography* 28 (3): 164–73.
Jones, T. 2012. 'Indonesian cultural policy in the reform era'. *Indonesia* 93: 147–76.
Kamariotou, V., M. Kamariotou and F. Kitsios. 2021. 'Strategic planning for virtual exhibitions and visitors' experience: A multidisciplinary approach for museums in the digital age'. *Digital Applications in Archaeology and Cultural Heritage* 21: e00183. https://doi.org/10.1016/j.daach.2021.e00183.
Kawashima, N. 1998. 'Knowing the public. A review of museum marketing literature and research'. *Museum Management and Curatorship* 17 (1): 21–39.
Kitley, P. 2014. *Television, Nation, and Culture in Indonesia*. Athens: Ohio University Press.
Knott, E. 2015. 'Everyday nationalism: A review of the literature'. *Studies on National Movements (SNM)* 3. Accessed 10 September 2024. https://openjournals.ugent.be/snm/article/85398/galley/203296/view/.
Kreps, C. F. 1994. 'On Becoming "Museum-Minded": A study of museum development and the politics of culture in Indonesia'. PhD dissertation, University of Oregon.
Kusno, A. 2012. 'Rethinking the nation'. In *The SAGE Handbook of Architectural Theory*, edited by C. Greig Crysler, S. Cairns and H. Heynen, 213–30. London: Sage.
Lang, C., J. Reeve, and V. Woollard. 2006. *The Responsive Museum: Working with audiences in the twenty-first century*. Farnham, Surrey: Ashgate.
Lanir, J., T. Kuflik, E. Dim, A. J. Wecker and O. Stock. 2013. 'The influence of a location-aware mobile guide on museum visitors' behavior'. *Interacting with Computers* 25 (6): 443–60. https://doi.org/10.1093/iwc/iwt002.
Lefebvre, H., translated by D. Nicholson-Smith. 1991. *The Production of Space*. Oxford: Blackwell.
Lekakis, E. J. 2017. 'Banal nationalism and consumer activism: The case of #BoycottGermany'. In *Everyday Nationhood*, edited by Michael Sky and Marco Antonsich, 285–304. New York: Springer.
Loscher, G., V. Splitter and D. Seidl. 2019. 'Theodore Schatzki's practice theory and its implications for organization studies'. In *Management, Organizations and Contemporary Social Theory*, edited by Steward R. Clegg and Miguel Pina e Cunha, 115–34. London: Routledge.

McDaid, S., S. Filippini-Fantoni and M. Cock. 2011. 'Handheld handholding: Small-screen support for museum visitors'. Conference proceedings, *Electronic Visualisation and the Arts*, 6–8 July 2011, 11–16. Accessed 10 September 2024. https://www.scienceopen.com/hosted-document?doi=10.14236/ewic/EVA2011.4.

McGregor, K. E. 2003. 'Representing the Indonesian past: The national monument history museum from guided democracy to the new order'. *Indonesia* 75: 91–122.

McTavish, L. 2006. 'Visiting the virtual museum: Art and experience online'. In *New Museum Theory and Practice*, edited by Janet Marstine, 226–46. Hoboken: Wiley. https://doi.org/10.1002/9780470776230.ch9.

Mesquita, S. and M. J. Carneiro. 2016. 'Accessibility of European museums to visitors with visual impairments'. *Disability & Society* 31 (3): 373–88.

O'Reilly, T. 2007. 'What is Web 2.0: Design patterns and business models for the next generation of software'. *Communications & Strategies* 1 (First Quarter 2007): 17. Accessed 10 September 2024. https://ssrn.com/abstract=1008839.

Pierroux, P., M. Bäckström, B. Brenna, G. Gowlland and G. Ween. 2020. 'Museums as sites of participatory democracy and design'. In *A History of Participation in Museums and Archives: Traversing citizen science and citizen humanities*, edited by Per Hetland, Palmyre Pierroux and Line Esborg, 27–45. London: Routledge.

Pink, S., H. Horst, J. Postill, L. Hjorth, T. Lewis and J. Tacchi. 2016. *Digital Ethnography: Principles and practice*. London: Sage.

Porsché, Y. 2022. 'Tracing museum exhibition reviews: References, replies and translations between the museum space and the mass media'. *Discourse, Context & Media* 46: 100579. https://doi.org/10.1016/j.dcm.2022.100579.

Putri, A. K. 2019. 'Representasi orde baru di Museum Jawa Tengah Ranggawarsita tahun 1975–2009 [Representation of new order in Ranggawarsita Museum, 1975–2009]'. Institutional thesis. University of Indonesia, Faculty of Science and Culture.

Riddell, P. G. 2002. 'The diverse voices of political Islam in post-Suharto Indonesia'. *Islam and Christian–Muslim Relations* 13 (1): 65–84. https://doi.org/10.1080/09596410210299.

Russo, A. 2012. 'The rise of the "media museum": Creating interactive cultural experiences through social media'. In *Heritage and Social Media*, edited by Elisa Giaccardi, 163–76. London: Routledge.

Schatzki, T. R. 2005. 'Practice mind-ed orders'. In *The Practice Turn in Contemporary Theory*, edited by Karin Knorr Cetina, Theodore R. Schatzki and Eike von Savigny, 50–62. London: Routledge.

Schlesinger, P. 1987. 'On national identity: Some conceptions and misconceptions criticized'. *Social Science Information* 26 (2): 219–64.

Schroeder, M. 2007. 'The Humean theory of reasons'. In *Oxford Studies in Metaethics*, Volume 2, edited by Russ Shafer-Landau, 195–219. Oxford: Clarendon Press.

Schweibenz, W. 1998. 'The "virtual museum": New perspectives for museums to present objects and information using the internet as a knowledge base and communication system'. *Isi* 34: 185–200.

Setten, G. 2005. 'Farming the heritage: On the production and construction of a personal and practised landscape heritage'. *International Journal of Heritage Studies* 11 (1): 67–79.

Silberman, N. and M. Purser. 2012. 'Collective memory as affirmation: People-centered cultural heritage in a digital age'. In *Heritage and Social Media*, edited by Elisa Giaccardi, 13–29. London: Routledge.

Silverstone, R. 2012. 'The medium is the museum: On objects and logics in times and spaces'. In *Towards the Museum of the Future*, edited by R. Miles and L. Zava, 161–76. London: Routledge.

Simon, N. 2010. *The Participatory Museum*. Santa Cruz: Museum 2.0.

Smith, A. D. 2009. *Ethno-Symbolism and Nationalism: A cultural approach*. London: Routledge.

Takahashi, T. 2010. *Audience Studies: A Japanese perspective*. London: Routledge.

Tapsell, R. 2012. 'Old tricks in a new era: Self-censorship in Indonesian journalism'. *Asian Studies Review* 36 (2): 227–45.

Taum, Y. Y. 2015. 'Collective Indonesian memories of the 1965 tragedy during New Order Regime'. [conference paper]. *The 1965 Coup in Indonesia: Questions of Representation: Fifty Years Later* held at Universitas Sanata Dharma, Yogjakarta, 21–22 October 2015.

Toyoshima, F., A. Barton and J.-F. Ethier. 2022. 'Affordances and their ontological core'. *Applied Ontology* 17 (2): 285–320. https://doi.org/10.3233/AO-220267.

Treinen, H. 1993. 'What does the visitor want from a museum? Mass-media aspects of museology'. In *Museum Visitor Studies in the 90s*, edited by Sandra Bicknell and Graham Farmelo, 86–93. London: Science Museum.

Triastuti, E. 2013. 'Indonesian Women Blogging: From Serambi Mekka (Aceh) to Batawi (Jakarta) and in between'. PhD dissertation, University of Wollongong, Australia.

Triastuti, E. 2017. 'Indonesia's internet blueprint: Shifting experience in media culture'. In *The Routledge Companion to Global Internet Histories*, edited by Gerard Goggin and Mark McLelland, 479–89. London: Routledge.

Triastuti, E. and Rakmani, I. 2011. 'Cyber Taman Mini Indonesia Indah: Ethnicity and imagi-nation in blogging culture'. *Interworking Indonesia Journal* 3 (2): 5–11.

Van Klinken, G. 2002. 'Disintegrasi: Indonesia in the post-Suharto era'. Fitzroy, Victoria: Arena Printing and Publications Pry. Ltd. Accessed 10 September 2024. https://www.thefreelibrary.com/Disintegrasi%3a+Indonesia+in+the+post-Suharto+era.-a083447469.

Van Klinken, G. 2005. 'The battle for history after Suharto'. In *Beginning to Remember: The past in the Indonesian present*, edited by Mary S. Zurbuchen, 233–58. Seattle: University of Washington.

Waterton, E. 2010. 'The advent of digital technologies and the idea of community'. *Museum Management and Curatorship* 25 (1): 5–11.

Witcomb, A. and K. Message. 2013. 'Introduction – Museum Theory: an expanded field'. In *The International Handbooks of Museum Studies*, edited by S. Macdonald and H. R. Levy, 35–63. Chichester: Wiley.

Woolley, S., J. Mitchell, T. Collins, R. Rhodes, T. Rukasha, E. Gehlken, E. Ch'ng and A. Cooke. 2021. 'Virtual museum "takeouts" and DIY exhibitions – augmented reality apps for scholarship, citizen science and public engagement'. In *Digital Heritage. Progress in Cultural Heritage: Documentation, preservation, and protection. EuroMed 2020*, edited by M. Ioannides, E. Fink, L. Cantoni and E. Champion, 323–33. New York: Springer. https://doi.org/10.1007/978-3-030-73043-7_27.

10
Replicas and religious heritage in the ethnographic museum

Ferdinand de Jong

Struggling for recognition, the discipline of anthropology presented its scientific knowledge to the public in spectacles such as colonial exhibitions (Coombes 1994). In Imperial France, the Musée de l'Homme and the Musée d'Ethnographie du Trocadéro conveyed evolutionary theory, as proposed by anthropology, to the wider public (Dias 1991; De l'Estoile 2007; Conklin 2013). As a panoptical institution the museum made evolutionary schemes perceptible (Bennett 2004). In the age of empire, the museum thus assumed a central role in the organisation of a new temporal conscience in which 'collections of national materials were represented as the outcome and culmination of the universal story of civilisation's development' (Bennett 1995, 77). In this order of time, the material culture collected in the colonies was used to demonstrate the 'arrested development' of 'primitive' others that contributed to the making of a colonial governmentality (Bennett et al. 2017). Given this history, it is not surprising that the recently opened Museum of Black Civilisations in Dakar, Senegal, has sought to break with this European museological tradition and shuns the ethnographic register (Bocoum 2018a; 2018b; Bocoum and Ndiaye 2022; De Jong 2022).

After anthropology was exposed as the handmaiden of colonialism (Asad 1973), the history of the ethnographic museum and its practices of collecting became the object of critical examination (Stocking 1985). More recently, requests for the restitution of objects looted under colonialism have increased the pressure for change in the ethnographic museum, resulting in a demand for epistemic decolonisation of institutions deemed institutionally racist (Hicks 2020). In response to such calls, ethnographic museums have made an effort to address the concerns of source communities and involve them in museum work (Peers and

Brown 2003; Clifford 2013; Lonetree 2012; Stanley 2007). But questions about the future of the ethnographic museum remain (Von Oswald and Tinius 2020). For instance, to what extent can the ethnographic museum contribute to repairing the time of colonialism (De Jong 2022)?

Because anthropology has displayed an interest in the subject of religion since its foundation, the ethnographic museum exhibited material culture to represent the world's religions. In the debate on secularism and secularity as it emerged in the early 2000s, Asad (2003) argued that the secularisms as they had appeared in Europe and its colonies were specific social formations. This is relevant in relation to the conception of the museum as a secular institution. Because they received their mandate in the Enlightenment museums have usually embodied the spirit of secularism, even though a critical literature has recast the museum as a sacred space (Duncan 1995; Bouquet and Porto 2005; Paine 2013; Buggeln et al. 2014; Oliphant 2021; Machabée 2023). How do we resolve this paradox of the multiple affordances of the secular museum? One way forward, I suggest, is to understand the museum as existing not in a linear temporality of 'progressive' secularisation, but as an assemblage of entangled temporalities of the sacred and the secular (De Jong 2023). In such an assemblage, museums may decolonise the museum as secular institution by affording an alternative, religious future. In the case presented here, replicas of religious shrines in the museum play a central role in decolonising the gaze and creating religious affordances for the ethnographic museum.

The Jola Museum

The Jola Museum is located at the village square of Mlomp, one of several villages inhabited by the Jola population on the southern shore of the Casamance River in Senegal's southernmost Casamance region. If the visitor to Mlomp were not familiar with the museum's existence, she would easily overlook the tiny construction. Initially, the museum was made of a circular wall of wooden sticks and bamboo poles about a man's height, covered with palm tree leaves. Its entrance was marked by a bamboo door which gave access to a single room approximately six metres in diameter (Figure 10.1). It was a remarkable construction, but one that was never meant to be the museum's permanent form. As its curator told me at the time, he really envisaged his museum to be housed in a so-called 'impluvium house'. A vernacular form of Jola architecture, an impluvium house is a circular house of around

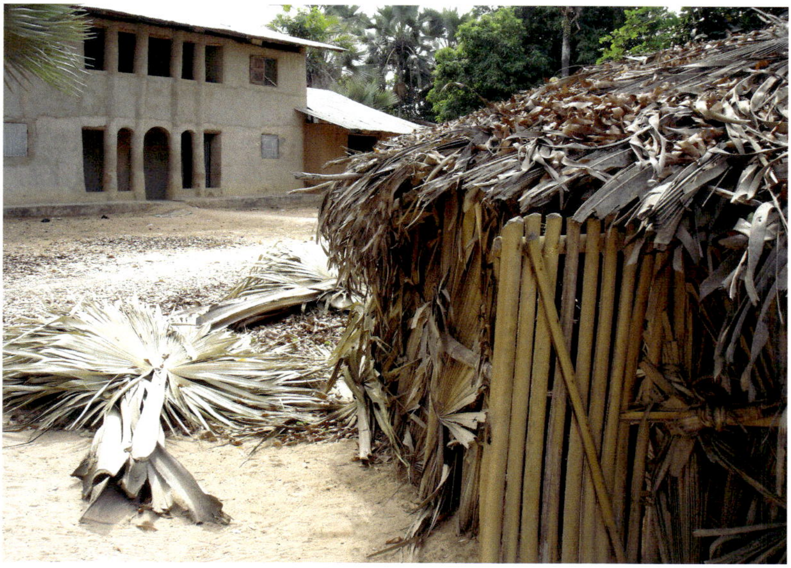

Figure 10.1 The Jola Museum in its original state, 2004 (the building in the foreground, on the right, situated in the square of Mlomp). Source: © Ferdinand de Jong, 2004.

15 metres in diameter to which access is provided by one door that opens to an inner court, around which various rooms are situated. An impluvium house can accommodate an entire extended family. In the ethnographic literature it is called an impluvium house because during the rainy season, water can be collected in the open space in the centre. In 2012, the curator had amassed sufficient funds to realise his dream and constructed an impluvium house to house the museum. The impluvium house lends itself well to hosting a museum as it is built around a central open space, lit by sunlight falling through the circular, open roof. As one of a small number of impluvium houses in the region, the Jola Museum draws upon the region's vernacular architectural heritage.

Tours through the museum are provided by the curator, Jules Sambou, who is the person who established the museum in 1992. A tour lasts for approximately 20 minutes, with the curator commenting on every object on display. Visitors are allowed to take pictures, and after the tour the curator requests that visitors make a donation. After my first visit to the museum in 2004 I befriended the curator, with whom I have remained in contact. Recently, we have been in touch quite often as the museum had crumbled due to severe rains (adobe constructions crumble for different reasons) and Sambou had the museum rebuilt. He regularly updated me on the progress of construction by sending me pictures on WhatsApp.

What exactly is on display in the Jola Museum? The collection comprises several objects that hang on the wall and a couple of constructions that sit on the floor. Most of the objects on display are household utensils, such as pottery and woven baskets. The full range of utensils includes agricultural tools, such as an old shovel used for wet-rice cultivation and a climbing-belt that men use for climbing palm trees to collect palm wine. Furthermore, the collection comprises weapons such as a bow and arrows, an ancient lance and a shield of hippopotamus skin. Thus, although a classification of the collection is not provided, alongside objects associated with subsistence farming there are objects related to hunting and warfare. In addition, there are a few curiosities: a parasol for the protection of a child carried on its mother's back and an earthenware pot used to store water (called 'a Jola fridge'). Finally, scattered throughout the museum are four contraptions labelled as *fétiches* (fetishes): there is one altar for the dead and one for confessions by the living, one fetish for marriage and fertility and one against the theft of rice, the latter usually installed in the rice fields. Initially, a little label was attached to all objects, explaining in French their use in daily life together with the original Jola gloss. For instance, a little container had a label stating, 'bucket, *ejund*, in which palm wine is served'. A woven basket had as explanation, 'basket, *elap*, to stock pounded rice (on the verge of disappearing)'. Indeed, most objects displayed in the museum are of local provenance and represent handcrafted objects no longer produced today. In his commentary, the curator often closes his comments on a particular object with that statement that 'this object is nowadays replaced by a plastic substitute which is bought on the market'.

The museum is charming owing to its simple construction and modest proportions. I, for one, was immediately seduced by this vernacular museum, wonderfully situated under a couple of huge silk-cotton trees. The objects on display are 'authentic' in the sense that most of them are truly old. All the exhibits are rare and some of them are outright spectacular (such as the shield made from hippopotamus skin – the hippo is since long extinct in this region). The commentary by the curator is comprehensive and understandable for those not initiated in Jola ethnography. And yet, my feelings toward the museum were ambivalent. I was aware that my view of the museum was informed by imperialist nostalgia (Rosaldo 1989). Salvage anthropology assumed that the ethnographic subject was destined to die out. For the making of its object, anthropology has always relied on the temporal distancing of its object. Ethnographic museums froze the colonial subject in an 'ethnographic present' that denied their coevalness (Fabian 1983). Yet, curator Jules Sambou never demonstrated any doubts about his ethnographic museum, suggesting that his appropriation of the genre of the ethnographic museum worked in ways that require us to think again. The remainder of this chapter aims to unpack how the museum may work to safeguard religious practices against a theory of progressive secularisation.

The contact zone

The existence of the Jola Museum raises interesting questions about the purpose of this ethnographic museum. To start, we may ask, why was the museum established in the first place? Jules Sambou, founder of the Jola Museum, was born in Mlomp in 1969 and went to school in nearby Oussouye. In 1985, his father, a palm wine tapper by profession, fell from a tree and could no longer afford to pay his son's school fees. Sambou had to break off his education and started guiding tourists around his village. He acquired a reputation for this and his name appeared in the popular French travel guidebook, *Le Guide du Routard* (Gloaguen 1986). As a tour guide, it occurred to him that most tourists were interested in shrines and old objects and he realised that it might be useful to create a place where these objects were conveniently brought together. Such practical considerations led to the creation of the museum. At least, this is what Sambou told me. On other occasions, he denied that the museum was his invention and declared it was God's gift, enabling him to make a living. Once the idea was born, Sambou discussed it with friends – trained seminarians – who encouraged him to pursue his project. His father also approved of the construction of a museum and advised him to consult the king (*oeyi*) of Mlomp, since Sambou wanted to establish the museum at the village's central square (Hutendukaa), which was the property of the king. The king gave his permission for Sambou to construct the museum at the square and Sambou and his friends started collecting objects and constructing the museum. It was officially opened at the beginning of 1992. Unfortunately, the timing was not auspicious: at that very moment the guerrilla war waged by the separatist movement MFDC reached its apogee and virtually no tourists visited the region, marked by civil war. Despite this ongoing conflict, the museum has stayed open to the public since 1992 and attracts a very modest number of visitors.

As a culturally and historically contingent practice of collecting, classifying and displaying objects, the museum is not part of the traditional Jola cultural repertoire. As far as the Jola have dedicated themselves to collecting, they have privileged rice, cattle and cloth, their traditional prestige goods. Every household has its own stock of rice – sometimes more than a decade old – stored in a granary and prevented from rotting by its strategic location above the household fire. Cattle are owned mostly by the family elder and are not kept for practical use in agricultural tasks – although their manure is used as fertiliser – but for slaughter during community-wide ceremonies of male initiation. Another occasion for the slaughter of cattle is the death of a male elder. His cattle are slaughtered as proof of his accumulated wealth and consumed in a potlatch in which all community members participate through the distribution of meat in various exchange networks. Cloth, too, is part of a moral economy of exchange and is distributed at the death of an elder. Interestingly, stocks of rice, cattle and cloth are never advertised – except at the time of their distribution – since accumulated wealth attracts accusations of witchcraft. Thus, the rice granary is locked and only accessible to its owner to prevent publicity about accumulated riches. The same applies to cattle; the herd usually is in the care of dispersed friends. Accumulated wealth is a well-guarded secret. This suggests that public accumulation and collection of valuables raises suspicion in the Jola moral economy.

Another reason why the Jola Museum parts from tradition is that it so ostensibly transfers cultural knowledge. Among the Jola knowledge is often defined as secret, especially the kind of knowledge transmitted during the initiation ceremony that all young men undergo (De Jong 2007a). As an institution for the transmission of knowledge, the Jola Museum is of course very different from the initiation ceremony and closer to forms of education introduced during colonial rule. In that sense, the museum represents an innovation in Jola instruction, comparable to other innovations introduced under colonialism. To sum up, the museum does not have roots in local forms of education, accumulation or archiving. However, in other ways the museum is very much a community enterprise. Most objects displayed in the museum were collected without much difficulty. Sambou obtained most of the objects from his maternal family where he has, as every Jola cousin does, a particular licence. In the Jola kinship system, sister's son (*assebul*) has the liberty to take anything from his mother's brother's compound, and Sambou exercised this privilege in his collecting of objects. Yet in some cases, negotiations were required to obtain an object. Sambou told me that he had been negotiating for years (offering large gifts, like pigs) to acquire a particularly old gourd that a barren woman (*anyalen*) had used in the traditional therapy that Jola women practise for restoring fertility. Despite these gifts, the woman did not come forth. In other cases, Sambou himself was hesitant in collecting objects. For example, a man informed Sambou that a pair of fetters could be found in the proximity of an abandoned shrine formerly used for chaining captives. He could freely collect these shackles, Sambou told me. At the time of my first visit, he was nonetheless waiting for a proper moment to collect them, fearing the disapproval of co-villagers.

Other objects were included in the museum collection without the risk of moral opprobrium. However, none of the objects in the collection was bought and the museum has virtually all objects on loan. The owners were reluctant to part with their heirlooms, some of which belonged to their forebears. Moreover, the inhabitants of Mlomp have a general reluctance towards monetised transactions. To express his gratitude for the loans, Sambou makes small annual gifts to the owners (in money or kind, a recurring transfer). In addition to the recompense offered to the loaners, the curator also makes regular offerings of palm wine to the king and his assistants. In exchange, they bless the museum and wish Sambou success in business. This allows me to conclude that the initiative to establish the museum was taken by a man who – in accordance with the courtesies required in a gerontocracy – consulted his male elders and the king, before undertaking action. Objects were collected in agreement with prevailing standards of ownership and exchange. Monetary payments were avoided as this kind of exchange was considered detrimental to appropriate standards of reciprocal behaviour, and the museum relies to a large extent on reciprocity, as indeed does the entire political economy of Mlomp.

Foreign visitors to the Jola Museum also like to see the museum as grounded in a local moral economy. After a tour through the museum, the visitors used to be confronted with a sign that said, 'the curator of the ethnographic museum of Mlomp requests your generosity'. Although gifts are frequently given in Mlomp's everyday life, we should not necessarily understand this request for a gift as forcibly emanating from a gift economy. Sambou wanted to maximise his profits and he was afraid that compulsory entrance fees might dissuade potential visitors from entering his museum. The curator calculated that gifts might provide him with a higher level of income than entrance fees and to this day, he continues this policy. His strategy should be situated within the cultural logic of tourism in which hosts graciously offer hospitality to their guests, sustaining a myth of mutual reciprocity that conceals the pecuniary nexus. The Jola Museum is a *contact zone* in which momentary reciprocity is created in a sphere of asymmetrical relations. As a local appropriation of the global museum practice, the Jola Museum is a space of somewhat hybrid transactions.

Appropriating anthropology, decolonising Catholicism

The Jola museum advertises itself as an ethnographic museum. A leaflet distributed by the museum states, 'Visit the ethnographic museum and discover Jola traditional culture, with its fetishes and utensils.' The exhibit offers what it promises: an overview of traditional Jola material culture. All the exhibits are locally sourced and the material culture exhibited in the museum privileges Mlomp's material culture, which is representative of Esulalu, identified by Baum (1999) as constituting a cultural Jola subgroup. Jola initiation masks like those that have found their way to European ethnographic museums are not part of the exhibit because they are not part of Mlomp's initiation ceremonies (Mark 1992; Mark, De Jong and Chupin 1998; De Jong 2007a).

Although the Jola Museum gives a partial view on Jola material culture, curator Sambou is aware of the selectiveness involved in the display. His ideas about traditional Jola culture are not idiosyncratic and are shared with his fellow Jola. His view on Jola material culture is derived, partly, from an ongoing conversation on Jola identity. This construction of Jola identity is much indebted to colonial ethnography and to the two-volume ethnography on the Jola by the eminent French ethnologist, Louis-Vincent Thomas (1959). His ethnography develops a typology of the Jola which demonstrates an uncanny resemblance to current Jola self-conceptions, and potentially even contributed to their making. Thomas praises 'the genuine Jola' (*le diola pur*) and presents the Muslim and Catholic Jola as culturally 'inauthentic'. According to this ethnography, which set the standards in the field for several decades, the 'genuine Jola' lived precisely in the Kasa region in which Mlomp is situated. Even today, many Jola feel that the authentic way of Jola life is preserved there. Although it is hard to prove that this notion of Jola authenticity is the result of colonial ethnography, corroboration for this thesis can be found in the fact that many supporters of the MFDC separatist movement were highly interested in the work of Louis-Vincent Thomas, which they considered an accurate description of the 'original' Jola way of life. Indeed, Thomas's work, as well as the historical study by

Christian Roche (1985) that heavily romanticises the resistance of the Jola against French colonisation, are well-known among Jola intellectuals (Foucher 2002b, 407). Even in the most remote villages of Lower Casamance one comes across copies of these books. In Mlomp, I was surprised to see how versatile its inhabitants are about Jola traditions, and many seem to have naturalised the role of the native informant (cf. Baum 1999, 11–19).

Why a Jola Museum should have been established can only be understood when we realise that alongside French ethnographers, Catholic missionaries have been involved in the Jola ethnogenesis. As in so many other instances of ethnogenesis, Jola identity is a recent invention (De Jong 1994). Contemporary notions of Jola authenticity are in fact derived from colonial anthropology and have been disseminated by Catholic missionaries in their instruction of Jola converts (Foucher 2003). With the colonial pacification of the Casamance region, a new frontier opened for the Catholic Church, which had experienced disappointing results in its missionary work in Senegal. Acknowledging that little success would be likely in the Senegalese colony, with the majority of the population already converted to Islam, the Casamance region offered perhaps a last chance to implant Christianity. Since the end of the nineteenth century, Holy Ghost fathers had slowly converted a significant minority of the Jola population on the southern shore of the Casamance River (those on the northern shore converted to Islam). While the Jola welcomed the missionaries as providers of education – many primary schools were established in the area – the majority of Jola found it hard to abandon their shrines as a condition to their conversion. As a result, their conversion was in many cases incomplete and quite a few Jola returned to their traditional ways (Baum 1990; Foucher 2002a). Others successfully hybridised traditional and Catholic practices in a new syncretism.

As a result of the input of African clergy attending the Second Vatican Council (1962–5), the Church showed a more accommodating attitude towards local religious practices and its 'inculturation' policy led to the Africanisation of theology and liturgy (Foster 2019). Although it was not framed this way, one might argue that 'inculturation' amounted to a 'decolonisation' of Catholicism. From the 1960s onwards, the missionaries encouraged Jola Catholics to take part in traditional rituals such as funerary rites (Baum 1990). The Catholic liturgy was adapted too, incorporating elements of Jola song and dance. While the Catholic missionaries had always relied on colonial ethnographers for their understanding of the Jola, an increasing number of Jola priests now dedicated themselves to the study of Jola cosmology to develop a Jola theology compatible with Catholic theology (Foucher 2003, 26). These auto-ethnographies contributed to the dissemination of a theology that was better adapted to the cosmology of Jola converts. Made the object of study by local priests, tradition offered a sense of dignity to its practitioners. The ethnographic knowledge thus produced has contributed to the making of Jola nationalism, particularly among Jola intellectuals living in exile (Foucher 2003; cf. Lambert 2002).

The production of an archetypal image of the Jola as hardworking, self-subsistent rice farmers with a sense of independence and strong attachment to their spirit shrines was subsequently appropriated by the Jola themselves. The Jola increasingly relied on their traditional rituals in their self-definition, especially in a context of globalisation. Today, diasporic Jola of both Muslim and Catholic persuasion return to their native villages to have their sons initiated in village-wide ceremonies, and participate in cultural festivals (*journées culturelles*) at which masquerades and theatrical productions on traditional life are performed (De Jong 2007a). This embrace of a reified tradition not only mitigates the alienation experienced by Jola living in diaspora but has been turned into a major tourist attraction as well. Whereas tourists visit *la petite côte* to enjoy the Senegalese beaches, the Lower Casamance region has become the primary destination for an experience of 'traditional culture'. Very important in that respect has been the development of a new form of tourism that offers accommodation in Jola villages. Small hostels, often constructed in ways reminiscent of ancient Jola architecture and run by village collectives, enable off-the-beaten-track tourists to stay overnight in villages and observe 'traditional' Jola life, as if they were anthropologists. Quotidian life is made the object of the tourist gaze: attractions on offer consist of dances, masquerades, wrestling matches and the Catholic mass said in Jola (Foucher 2002b; De Jong 2007a). Since 1992, the Jola Museum has been included in these tours as one of these attractions.

The production of a Jola cultural specificity – either for theological or touristic purposes – has been inspired to a large extent by colonial anthropology. The Jola have always attracted ethnologists and anthropologists interested in radical alterity. In this respect, Foucher has correctly characterised the anthropology of the Jola as preoccupied with Jola traditional religion (Foucher 2005, 367). The first French monographs (Thomas 1959; Girard 1969) included photographs of spirit shrines, representing them as exemplifying a cosmology peculiar to the Jola. American scholars too have focused their research on spirit shrines and have demonstrated how these shrines have been pivotal in regulating Jola social and economic life (Mark 1985; Linares 1992; Baum 1999; 2016). The Jola have familiarised themselves with these works that they believe represent their 'quintessential' selves. It is through such ethnography that a Jola identity has been fashioned (Foucher 2011).

The Jola Museum is a space in which this academic knowledge turned into popular knowledge, based on a vision authorised by colonial anthropology and the Catholic Church, appropriated by the Jola population, and commodified for the tourist industry. Although these processes involved a range of international institutions and global historical processes, they can be seen at work in the Jola Museum and the life of its curator. Raised in a Catholic family in the 1970s, Sambou himself has grown up with 'inculturation' liturgy and theology. Much of his social life is in Catholic circles, and his involvement in the Jola Museum too seems to have been decisively influenced by the Church. One of the Piarist Fathers based in the region assisted Sambou with the establishment of the museum. Being Catalan, the Father felt that both Catalan and Jola cultures were under threat; he was therefore sympathetic towards the establishment of a museum of Jola culture. Cultural nationalisms met and reinforced each other around a practice of curation. Supportive of a museum that could 'salvage' Jola culture, he assisted Sambou financially. To inform Sambou of the academic knowledge on the Jola, he lent him his copy of Louis-Vincent Thomas's *Les Diola* (1959). Clearly, the Jola Museum is a belated consequence of colonial anthropology that to this day provides an authoritative context to the Jola Museum. The four 'fetishes' already mentioned, placed around the museum, provide material testimonies to the Catholic acknowledgement of traditional Jola religion. In sum, inspired by the ethnographic museums that emerged in late nineteenth-century Europe to demonstrate the life of colonised populations, the Jola Museum of Mlomp has appropriated the format as a way of imagining the Jola nation. What at first looks like a commercial enterprise to tap into a cash flow generated by tourism appears on closer inspection to be a carefully arranged selection of Jola material culture, its historicity surreptitiously concealed behind a carefully crafted 'ethnographic present'.

Talking points

Oscillating between subject and object of ethnography, the Jola Museum authorises views on the Jola, but also raises questions about the authority an auto-ethnographic museum can assume. Through their association with academic disciplines, museums have historically mediated the production of knowledge. Survey museums in European capitals have assumed an authority that privileged Europe as the subject of the world. Visits to these museums have been understood as civic rituals that enabled citizens to acknowledge the imperial nation-state as harbingers of civilisation. Obviously, the Jola Museum is not comparable to such national, encyclopaedic museums. As an ethnic museum situated in the periphery of the national territory, the Jola Museum does not authorise the Senegalese state. But which narrative does it authorise?

Even if the structure of the Jola Museum evokes the most accomplished genre of Jola architecture, its claim to antiquity is incomparable to those European museums that deliberately recall the ceremonial architecture of Greek temples (Duncan 1995). Unlike the Louvre and other national survey museums situated in the epicentres of European metropoles, the Jola Museum has no corridors that enable ceremonial procession through a succession of rooms. Whatever narrative is told here, it cannot be one of a long evolution towards civilisation. Instead of marble columns and the occasional *trompe l'oeil*, the Jola Museum is made of adobe clay and palm leaves. If national survey museums were built for eternity, the Jola Museum requires the annual replacement of its decaying materials and reminds one instead of the perishable nature of human achievements. Sitting under huge cotton trees that dwarf the museum, the Jola Museum does not intend to imitate a European model of museums. It differs in its claim to authority by deliberately invoking the renewable organics of a vernacular architecture that the European imagination has always associated with 'primitive' society. The radical alterity of the Jola Museum provincialises the European gaze trained by the encyclopaedic museum.

The anthropological literature authorises the museum to those initiated in Jola ethnography, but of course most tourists are unfamiliar with this form of esoteric knowledge. As curator, Sambou himself never references anthropological studies. In contrast, he references the various local authorities that he consulted before constructing the museum. He tells the visitors that the king had given his consent to the construction of the museum and that when the museum was subsequently vandalised, the king had convened a meeting to tell the assembled villagers that he approved of the museum. This allusion to the authority of the king clearly frames the museum as subjected to a local, native authority (see Figure 10.2). This form of authorisation also

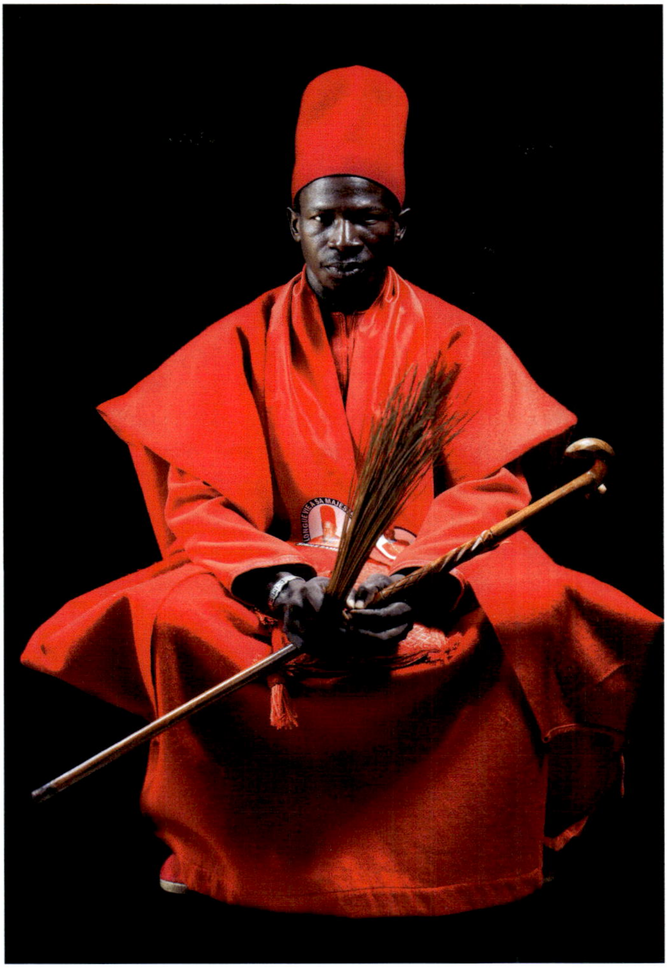

Figure 10.2 The present king (*oeyi*) of Mlomp, Sibilé Sambou.
Source: © Matar Ndour (2024), used by permission.

appears in the way in which the French travel guidebook *Le Guide du Routard* advertises the Jola museum:

> Not to be missed for a number of reasons: the museum is modest but with a 20-minute tour that is well-commented you will appreciate much better the villagers that constantly follow you and you will help to safeguard a culture that merits salvation; you can talk to Jules who will voluntarily explain how he got the idea of the museum and how he obtained authorisations and advice from

> the elders, the king (a religious function) and the *chef du village*. Finally, if your interest in Jola tradition is sincere, you will be invited to drink palm wine, of which this region produces a great deal. Jules will even explain the technology used to obtain the sap of the palm tree without killing it; his father was a famous palm wine tapper. (Gloaguen 1986, 206 [author's translation])

The travel guide presents Jules Sambou as the founder of the museum and his museum is presented as authorised by the elders, the king and the *chef du village*. Interestingly, while the travel guide authorises the museum through references to 'traditional' authorities, Sambou himself, in conversations with me, also authorised his museum through references to the same guide. Sambou also owns a copy of the *Directory of Museums in West Africa*, published by the West African Museums Programme (2000), which includes a description of his museum. The museum is authorised in different registers. The process of authorisation is thus based on an acknowledgement of 'traditional' authorities and various documents, providing an intertextual authorisation that confirms Derrida's observations that '[b]y incorporating the knowledge deployed in reference to it, the archive augments itself, engrosses itself, it gains in *auctoritas*' (Derrida 1996, 68).

Concealment and revelation

If the museum is one of the most important institutions in the production of modernity, the discourse of modernity itself has been dependent on 'magic' as exemplifying modernity's alterity (Meyer and Pels 2003). In an elucidating essay, Peter Pels (2003, 5) says that 'anthropology, more than any other scholarly discourse on magic, was responsible for the interpretation of magic as an antithesis of modernity and the production of the peculiar ambiguity and entanglement of magic and modernity'. In this discourse, a range of ambiguities developed with reference to fetishism and the fetish. William Pietz has provided us with a cultural history of the 'fetish', the object that European travellers in sixteenth-century West Africa thought Africans attributed unparalleled power and value to. Pietz (1988) demonstrates how descriptions of the 'fetish' entered seventeenth-century Dutch proto-anthropology, thereby laying the foundations for what became a powerful trope in the European imagination confounding subjects and objects in the mercantile world of the slave trade (Matory 2018). As Robert Baum (1999) has amply demonstrated, the shrines of the Esulalu Jola have been sites at which the trade in human beings was authorised. Replicas of shrines made of clay, sticks and skulls – leftovers of sacrifices – are a major attraction in the museum. So, what kind of object lessons does the museum tell us about its 'fetishes'?

Shortly after the museum was opened, it was vandalised by an inhabitant of the village. Someone made a hole in the wall and took away the sign 'Musée de la culture diola'. After this event, Sambou sought the help of the king to address this act of disapproval. Although the culprit was never identified, Sambou has some clues as to why the museum was vandalised. The reason, he assumed, must have to do with the four 'fetishes'. These fetishes are replicas of four different types of shrines as they exist in Mlomp. These shrines have been an object of academic interest ever since scholars have studied the Jola, as we already said above. In the village, spirit shrines can only be accessed by the members of well-defined social categories, for instance by initiated males, or women who have borne children. Access to the shrines and the transactions with them are subject to taboos which are usually not negotiable. Knowledge about the shrines and how to approach the spirits through them is monopolised by their priests. Knowledge of the shrines is esoteric and not made readily available (Baum 1999, 19).

It is correct to qualify the habitus of villagers towards the shrines as one of secrecy which, I have demonstrated elsewhere (De Jong 2007a), is a mode of performance characteristic of Jola ritual and religion. In Jola religion, everything sacred is necessarily cast as secret, and vice versa. The appropriate behaviour towards spirit shrines by Jola who have not been initiated in them would be one of avoidance. Let me illustrate this with a small sketch. One day, Sambou and I strolled through the village on a path that led past a shrine that he wanted to show me, as we had discussed it in our previous conversations. When we approached the shrine, Sambou simply walked past it. He encouraged me to take in the shrine from a distance: I was not allowed to stop and stand still, let alone take a picture. Whenever Sambou drew attention to sites he framed as 'secret', he simultaneously invoked taboos on the exercise of the gaze. In ways that seemed to replicate local ways of enforcing secrecy, Sambou enhanced the sacredness of secret sites by subjecting them to restrictions to the gaze.

Sambou is well trained in the ways in which the gaze contributes to the aura of shrines and has developed some expertise in managing the gaze of others. His decision to create the Jola Museum was, to some extent, a response to the fact that the villagers experienced the tourists as intruders who exercised their unrestricted gaze in the village. Clearly, there was a tension between the villagers' attitude toward their shrines and the interest that tourists took in them. What particularly annoyed the villagers was the tourists' habit of photographing everything, including shrines. Enabling the tourists to gaze on replica shrines exhibited in the Jola Museum meant their gaze was diverted from the shrines in the village. In our conversations, Sambou explicitly addressed the problematic nature of the gaze on sacra and informed me that this was one of the reasons he had founded the museum in the first place.

I was not the only person with whom he shared this narrative. When I attended a museum tour, a tour guide introduced the museum to her (French) audience by stating that 'Sambou was the founder of this little museum, which enables him to make a modest living which is, after all, much better than begging', going on to explain that the museum exhibits these fetishes so that tourists can be informed about them. 'After all', she went on, 'visiting fetishes at their sites in the village could possibly embarrass the villagers; some aspects of these shrines are actually secret'. With these words the tour guide framed knowledge about the shrines as secret and presented the museum as a legitimate site for their revelation. Although the shrines on display in the Jola Museum are replicas, the museum clearly buys into the dualism of secrecy and revelation to suggest that a visit to the museum initiates the visitor into the secrets of Jola society. As tourists *experience* the secrecy to be observed vis-à-vis shrines within the context of the museum, secrecy is thus made a tourist attraction.

Shrines and their replicas

There is no doubt that an urge to get hold of auratic objects drives tourists to photograph the Jola shrines. Informed of the ethnography of shrines by the curator, they are allowed to take pictures of their replicas in the museum. While taking pictures, the visitors are aware that they cannot take such pictures outside the museum. One may understand the staging of sacra as part of an interplay of the visible and the invisible whereby these sacra are made available in the museum in a way they are not in the village (cf. Pomian 1994). This is precisely what is at stake in the Jola Museum, which offers photo opportunities not provided elsewhere. As I argued elsewhere (De Jong 2007b; 2013), heritagisation requires the sacred to be made visible and subjects it to a regime of visibility that allows tourists to photograph sacra. The shrines on display in the

Jola Museum are replicas and the authorised focus of an experience of simulated secrecy.

Although most villagers never visit the museum and must guess what the tourists are told, the secrecy practised in the museum involves them too. The practice of secrecy in the village serves to demarcate social boundaries in the sphere of the 'real', but the museum replicates the practice in the sphere of 'representation'. This raises interesting questions about secrecy, the distinction between reality and its representation, and the reproduction of aura. Walter Benjamin (Taussig 1999, 216) suggests that auratic objects produce in their spectators a sense of distance; they require this distance to maintain their auratic quality. To frame the museum as a site for the transmission of secret knowledge clearly contributes to the replication of aura for the replica shrines in the museum. Indeed, the policy adopted by the museum affords replica shrines to produce their own auras (cf. Foster and Jones 2019). Replicas therefore take on the burden of the aura and replicate the sacred in the context of the museum.

Nonetheless, the distinction between replica and reality is the cause of some controversy surrounding the Jola Museum. Certainly, Sambou makes it clear to the visitors that the 'shrines' in his museum are replicas. Most tourists are familiar with the notion of replicas and make a distinction between the 'representations' in the museum and the 'originals' in the village. However, this distinction was not accepted as self-evident by one of the Senegalese tour guides who occasionally took tourists to the Jola Museum. In the conversation I had with him, he assumed that European visitors might mistakenly believe that what they saw was 'the real thing'. Ironically, the accusation of superstition Europeans historically levelled against African believers in 'fetishes' is here inverted and turned toward Europeans. It confirms Gwyneira Isaacs's (2011) observation that the distinction between 'originals' and 'copies' should not be assumed to be 'natural'. Blurring the boundaries between 'real' and 'replica', the replicas of the Jola Museum produce epistemological doubt.

In the light of the above, I posit that the vandalising of the museum had its origins in confusion around the replicas, and the natural distinction the museum assumed between the real and the replica. Analogous to Taussig's interpretation of the sacrilege of an Australian artwork, we might suggest that the replicas came alive 'in a spurt of mimetic efflorescence' (Taussig 1999, 28). This moved the vandal to deface the museum. An intervention by the king of Mlomp calmed down the villagers and re-assured them of the bona fide nature of replication. At first perceived as a place of exposure, the Jola Museum was subsequently understood as a site of replication. Today, the villagers understand the sort of thing Sambou is doing, and so do the tourists. The museum has produced a space of replication whereby the villagers remain enchanted by the originals, while the tourists are enchanted by the replicas and can photograph them in good faith that they are not violating the originals. By keeping the replicas shrouded from the gaze of the villagers, the Jola Museum has maintained that verge whereby the secret is not destroyed through exposure, 'but subject to a quite different sort of revelation that does justice to it' (Walter Benjamin quoted in Taussig 1999, 3).

Conclusion

Timothy Mitchell (1991, 32) argues that the exhibition has created a world divided in a realm of mere representations and a realm of the 'real'. While the first is to be found within the space of the museum, the latter is to be found in the realm of the 'real', the 'outside' and the 'original', which are constructions just as much as is the realm of 'representation'. One can no longer assume that the distinction between 'originals' and 'replicas' is stable. In fact, by performing secrecy at the Jola Museum, the tourists blur the distinction between the secret and its simulacrum in this contact zone of transcultural performance (De Jong 2007a). Even though the distinction between originals and representations made by the Jola Museum is not entirely efficacious (cf. Meyer 2015, 273), it is by redirecting the gaze that the aura of the 'original' shrines is maintained.

The distinction between 'originals' and 'replicas' might be one of the distinctions that make up our belief in the modern (Latour 1993). What obfuscates the distinction between these two realms, is that secrecy is enacted vis-à-vis the 'original' as well as the 'replicas'. Instead of secreting the 'original' from the 'replica', secrecy binds them in a set of ambiguous relations that resonate with and replicate the historical context of interactions between Europeans and West Africans around objects attributed power and value. Shrines, Jola religion, and Catholicism constitute an entangled set of practices that can be understood as a replication of religion (Coleman 2023).

Replicas, it has recently been argued, produce their own authenticity. The regime of visibility surrounding the shrines in the village is doubled by a regime in which the museum makes shrines visible. Secrecy is replicated in the Jola Museum and renders the museum a site for the replication of religion. In the last chapter of *Magic and Modernity*, Michael Taussig (2003) makes a compelling argument that the magic that anthropology's founding fathers attributed to modernity's others was part and parcel of the production of their foundational texts on magic. James Clifford (1988) made a similar argument with regard to secrecy and initiation in the texts of the French ethnologists. Clearly, modernity hinges on a play of concealment and revelation that the Jola Museum, as a site of anthropological knowledge production, replicates and reproduces for the future.[1]

Note

1 Fieldwork for this paper was conducted in 1997–8 and in many subsequent visits to the Casamance region of Senegal, most recently in 2023. I would like to thank Jules Sambou, Curator of the Jola Museum, for his kind collaboration and friendship throughout this period. I also thank the editor of this volume, Nick Stanley, for his constructive comments on previous drafts of this chapter. Finally, I thank Mater Ndour for his permission to reproduce his photograph of king Sibilé Sambou. An earlier version of this chapter was published in the *Journal of Museum Ethnography* in 2002 (De Jong 2022).

References

Asad, Talal (ed.). 1973. *Anthropology and the Colonial Encounter*. London: Ithaca Press.

Asad, Talal. 2003. *Formations of the Secular: Christianity, Islam, modernity*. Stanford: Stanford University Press.

Baum, Robert. 1990. 'The emergence of a Diola Christianity'. *Africa* 60 (3): 370–98.

Baum, Robert. 1999. *Shrines of the Slave Trade: Diola religion and society in precolonial Senegambia*. Oxford: Oxford University Press.

Baum, Robert. 2016. *West Africa's Women of God: Alinesitoué and the Diola prophetic tradition*. Bloomington and Indianapolis: Indiana University Press.

Bennett, Tony. 1995. *The Birth of the Museum: History, theory, politics*. London: Routledge.

Bennett, Tony. 2004. *Pasts Beyond Memory: Evolution, museums, colonialism*. London: Routledge.

Bennett, Tony, Fiona Cameron, Nélia Dias, Ben Dibley, Rodney Harrison, Ira Jacknis and Conal McCarthy. 2017. *Collecting, Ordering, Governing: Anthropology, museums, and liberal government*. Durham, NC: Duke University Press.

Bocoum, Hamady. 2018a: 'Les grandes lignes du projet de Musée des Civilisations Noires à Dakar [The great vision of the Museum of Black Civilizations project in Dakar]'. In *Vers le Musée Africain du XXIe Siècle: Ouverture et cooperation* [Towards the African Museum of the Twenty-First Century: Openness and cooperation], edited by H. Bocoum, E. Féau and C. Crémière, 120–35. Le Havre: MkF Éditions.

Bocoum, Hamady. 2018b. 'Le Musée des Civilisations Noires: Une vision d'avenir. Interview du Hamady Bocoum'. *Présence Africaine* 197: 183–94.

Bocoum, Hamady and Malick Ndiaye. 2022. 'Le Musée des Civilisations Noires: A continuous creation of humanity'. In *National Museums in Africa: Identity, history, and politics*, edited by Raymond Silverman, Georges Abungu and Peter Probst, 127–38. London and New York: Routledge.

Bouquet, Mary and Nuno Porto (eds). 2005. *Science, Magic and Religion: The ritual processes of museum magic*. New York: Berghahn.

Buggeln, Gretchen, Chrispin Paine and S. Brent Plate (eds). 2014. *Religion in Museums: Global and multidisciplinary perspectives*. London: Bloomsbury.

Clifford, James. 1988. *The Predicament of Culture: Twentieth-century ethnography, literature, and art*. Cambridge, MA: Harvard University Press.
Clifford, James. 2013. *Returns: Becoming indigenous in the twenty-first century*. Cambridge, MA: Harvard University Press.
Coleman, Simon. 2023. 'Re-living religion: Ritual and heritage in English cathedrals'. In *The Future of Religious Heritage: Entangled temporalities of the sacred and the secular*, edited by F. de Jong and J. Mapril, 44–64. London: Routledge.
Conklin, A. L. 2013. *In the Museum of Man: Race, anthropology, and empire in France, 1850–1950*. Ithaca: Columbia University Press.
Coombes, A. E. 1994. *Reinventing Africa: Museums, material culture and popular imagination*. New Haven: Yale University Press.
De Jong, Ferdinand. 1994. 'Jola inventing their past and future'. *Leidschrift* 10 (2): 15–35.
De Jong, Ferdinand. 2002. 'The Jola Museum: An appropriation of museum practice'. *Journal of Museum Ethnography* 14: 4–13.
De Jong, Ferdinand. 2007a. *Masquerades of Modernity: Power and secrecy in Casamance, Senegal*. Edinburgh: Edinburgh University Press for the International African Institute, London.
De Jong, Ferdinand. 2007b. 'A masterpiece of masquerading: Contradictions of conservation in intangible heritage'. In *Reclaiming Heritage: Alternative imaginaries of memory in West Africa*, edited by F. de Jong and M. Rowlands, 161–84. Walnut Creek: Left Coast Press.
De Jong, Ferdinand. 2013. 'Le secret exposé: Révélation et reconnaissance d'un patrimoine immatériel'. *Gradhiva: Revue d'anthropologie et d'histoire des arts* 18: 98–123.
De Jong, Ferdinand. 2022. *Decolonizing Heritage: Time to repair in Senegal*. Cambridge: Cambridge University Press.
De Jong, Ferdinand. 2023. 'Introduction: Temporalities of renewal in religious heritage'. In *The Future of Religious Heritage: Entangled temporalities of the sacred and the secular*, edited by F. de Jong and J. Mapril, 1–21. London: Routledge.
De L'Estoile, Benoît. 2007. *Le Goût des Autres: De l'exposition coloniale aux arts premiers*. Paris: Flammarion.
Derrida, Jacques. 1996. *Archive Fever: A Freudian impression*. Chicago: Chicago University Press.
Dias, Nélia. 1991. *Le Musée d'Ethnographie du Trocadéro (1978–1908): Anthropologie et muséologie en France*. Paris: Éditions du Centre National de Recherche Scientifique.
Duncan, C. 1995. *Civilizing Rituals: Inside public art museums*. London: Routledge.
Fabian, Johannes. 1983. *Time and the Other: How anthropology makes its object*. New York: Columbia University Press.
Foster, Elizabeth A. 2019. *African Catholic: Decolonization and the transformation of the church*. Cambridge, MA: Harvard University Press.
Foster, Sally M. and Siân Jones. 2019. 'The untold heritage value and significance of replicas'. *Conservation and Management of Archaeological Sites* 21 (1): 1–24.
Foucher, V. 2002a. 'Cheated Pilgrims: Education, migration and the birth of Casamançais nationalism (Senegal)'. PhD thesis, SOAS University of London.
Foucher, V. 2002b. 'Les "évolués", la migration, l'école: Pour une nouvelle interpretation de la naissance du nationalisme casamançais [The 'evolved', migration, school: Towards a new interpretation of the birth of Casamance nationalism]'. In *Le Sénégal contemporain* [Contemporary Senegal], edited by M.-C. Diop, 375–424. Paris: Karthala.
Foucher, V. 2003. 'Church and nation: The Catholic contribution to war and peace in Casamance (Senegal)'. *Social Sciences & Missions* 13: 11–40.
Foucher, V. 2005. 'La guerre des dieux? Religions et séparatisme en Casamance [The war of the gods? Religions and separatism in Casamance]'. *Canadian Journal of African Studies* 39 (2): 361–88.
Foucher, V. 2011. 'On the matter (and materiality) of the nation: Interpreting Casamance's unresolved separatist struggle'. *Studies in Ethnicity and Nationalism* 11 (1): 82–103.
Girard, Jean. 1969. *Genèse du pouvoir charismatique en Basse Casamance (Sénégal)* [Genesis of charismatic power in Lower Casamance (Senegal)]. Dakar: IFAN.
Gloaguen, Phillipe. 1986. *Le Guide du Routard – Afrique Noire 1985–1986* [Rough Guide – Sub-Saharan Africa 1985–1986]. London: Hachette.
Hicks, Dan. 2020. *The Brutish Museums: The Benin Bronzes, colonial violence and cultural restitution*. London: Pluto Press.
Isaac, Gwyneira. 2011. 'Whose idea was this? Museums, replicas, and the reproduction of knowledge'. *Current Anthropology* 52 (2): 211–33.

Lambert, Michael. 2002. *Longing for Exile: Migration and the making of a translocal community in Senegal, West Africa*. Portsmouth: Heinemann.

Latour, Bruno. 1993. *We Have Never Been Modern*. Cambridge, MA: Harvard University Press.

Linares, Olga. 1992. *Power, Prayer and Production: The Jola of Casamance*. Cambridge: Cambridge University Press.

Lonetree, A. 2012. *Decolonizing Museums: Representing Native America in national and tribal museums*. Chapel Hill: University of Carolina Press.

Machabée, Stéphanie. 2023. 'A museum made mosque: Postsecularism and the case of Istanbul's Hagia Sophia'. In *The Future of Religious Heritage: Temporalities of the sacred and the secular*, edited by F. de Jong and J. Mapril, 25–43. London: Routledge.

Mark, Peter. 1985. *A Cultural, Economic and Religious History of the Basse Casamance Since 1500*. Stuttgart: Steiner Verlag/Frobenius-Institut.

Mark, Peter. 1992. *The Wild Bull and the Sacred Forest: Form, meaning, and change in Senegambian initiation masks*. New York: Cambridge University Press.

Mark, Peter, F. de Jong, and C. Chupin. 1998. 'Ritual and masking traditions in Jola men's initiation'. *African Arts* 31 (1): 36–47, 94–5.

Matory, J. Lorand. 2018. *The Fetish Revisited: Marx, Freud and the gods Black people make*. Durham, NC: Duke University Press.

Meyer, Birgit. 2015. *Sensational Movies: Video, vision, and Christianity in Ghana*. Berkeley: University of California Press.

Meyer, Birgit and Peter Pels (eds). 2003. *Magic and Modernity: Interfaces of revelation and concealment*. Stanford: Stanford University Press.

Mitchell, Timothy. 1991. *Colonising Egypt*. Berkeley: University of California Press.

Oliphant, Elayne. 2021. *The Privilege of Being Banal: Art, secularism, and Catholicism in Paris*. Chicago: University of Chicago Press.

Paine, Crispin. 2013. *Religious Objects in Museums: Private lives and public duties*. London: Bloomsbury.

Peers, Laura and Alison K. Brown (eds). 2003. *Museums and Source Communities: A Routledge reader*. London: Routledge.

Pels, Peter. 2003. 'Introduction: Magic and modernity'. In *Magic and Modernity: Interfaces of revelation and concealment*, edited by Birgit Meyer and Peter Pels, 1–38. Stanford: Stanford University Press.

Pietz, W. 1988. 'The problem of the fetish, IIIa: Bosman's Guinea and the Enlightenment theory of fetishism'. *Res* 16: 105–23.

Pomian, Krzysztof. 1994. 'The collection: between the visible and the invisible'. In *Interpreting Objects and Collections*, edited by Susan M. Pearce, 160–74. London: Routledge.

Roche, Christian. 1985. *Histoire de la Casamance: Conquête et résistance: 1850–1920* [History of Casamance: Conquest and resistance]. Paris: Karthala (reprint of orig. 1976 edition).

Rosaldo, Renato. 1989. 'Imperialist nostalgia'. *Representations* 26 (spring): 10–22.

Stanley, Nick (ed.). 2007. *The Future of Indigenous Museums: Perspectives from the southwest Pacific*. New York: Berghahn.

Stocking, George W. Jr (ed.). 1985. *Objects and Others: Essays on museums and material culture*. Madison: University of Wisconsin Press.

Taussig, Michael. 1999. *Defacement: Public secrecy and the labor of the negative*. Stanford: Stanford University Press.

Taussig, Michael. 2003. 'Viscerality, faith, and skepticism: Another theory of magic'. In *Magic and Modernity: Interfaces of revelation and concealment*, edited by Birgit Meyer and Peter Pels, 272–306. Stanford: Stanford University Press.

Thomas, Louis-Vincent. 1959. *Les Diola: Essai d'analyse fonctionnelle sur une population de Basse-Casamance* [The Diola: Functional analysis of a population of Basse-Casamance] (2 vols). Dakar: IFAN.

Von Oswald, Margareta and Jonas Tinius (eds). 2020. *Across Anthropology: Troubling colonial legacies, museums, and the curatorial*. Leuven: Leuven University Press.

West African Museums Programme. 2000. *Directory of Museums in West Africa*. Dakar: WAMP.

11

Entangled knowledges: re-indigenising biocultural collections at National Museums Scotland

Alison Clark, Shona Coyne, Alistair Paterson and Tiffany Shellam

In 2019, Yorta Yorta artist Treahna Hamm exhibited her artwork *A Yorta Yorta Person's Bush Medicine Kit* in an exhibition about First Australian Bush Medicine in Melbourne, London and Berlin (Healy 2018). The artwork, which replicated a European medicine kit, challenged the narrative put forward by European colonialists in Australia that the continent was empty of medicinal knowledge and practices. The bush medicine kit contained remedies that have been in use by First Australians for thousands of years and demonstrated the vital connection between people, Country (the term used by First Australian peoples to describe the connection between people, lands, seas and waterways, also encompassing language, law, cultural practices, spiritual beliefs and identity), and culture in Australia, a continent that is home to the oldest living biocultural knowledge on earth. From the 1700s onwards this knowledge was collected, categorised and removed from Country by European explorers, colonial officials, missionaries, anthropologists, zoologists, botanists and travellers.

The Enlightenment (1680–1820) significantly changed how Europeans viewed the world. Enlightenment thought and the expansion of Empire were intimately intertwined and with that came the development of the encyclopaedic or universal museum in which natural history and ethnography were exhibited side by side, with material understood as either a natural or artificial curiosity. European colonial expansion in the late 1800s to early 1900s was inextricably linked to the development and broadening of these early systems of taxonomy and with those distinct disciplines emerged. The emergence of these disciplines was linked to the development of practices of collecting, and with it discipline-specific museums (Haraway 1989; Stocking 1992; Thomas 1991; 1994). Guides and manuals for collecting were produced by societies and institutions keen to control the way that collections – with the future potential of becoming included in museums – were made, prepared and documented (Herschel 1849; British Association for the Advancement of Science 1874; Freshfield 1889). While shaping those collections that entered these museums, the guides also encouraged the formation of expansive collections across the disciplines. The problem was that when these expansive collections entered the museum they were often separated out into their disciplines, often losing the context that linked them (Alberti 2012). Collections such as those made by Scottish-born Assistant Commissary Officer Robert Neill (1801–1852) between 1841–5 in south-west Western Australia or those by Prussian-born entrepreneur Emile Clement (1844–1928) between 1896–1928 in north-west Western Australia while defined in Western taxonomic terms as containing both natural history and ethnography material are ultimately holistic snapshots of life in those places at that time (see Figure 11.1). Discussing Australia, artist Brian Martin describes Country and its materials as having ancestry and agency, writing that the removal of ancestors has 'deep spiritual and ontological ramifications for people and Place' (Martin 2023, 33). He further argues that 'dispossessing Place from materials is about removing and stripping their agency and subjectivity to create objects'

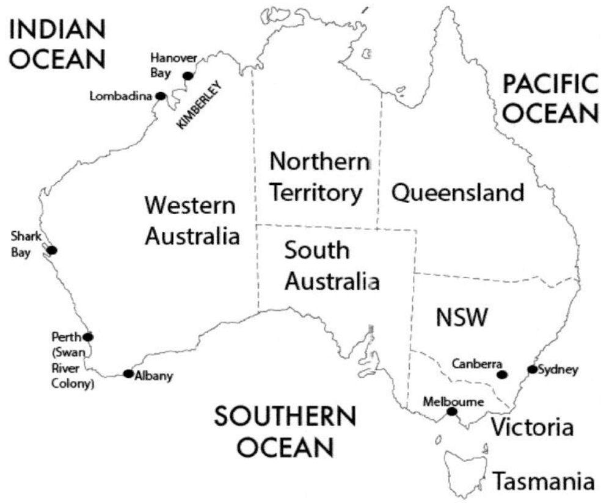

Figure 11.1 Map showing the collecting locations of Emile Clement and Robert Neill in Western Australia. Source: © Alistair Paterson 2024.

that are subject to dominant white epistemologies (33). In Australia, both Neill and Clement recorded indigenous knowledge about Country and, as we shall discuss, in many cases this indigenous knowledge was used to inform Western scientific knowledge demonstrating, as Treahna Hamm's artwork does, that the continent was not devoid of indigenous science. However, on their arrival into Europe and in compliance with Western taxonomic standards the collections became separated out into natural history and ethnography. Due to this dispersal and the dispersal of historical records, these collections and their associated indigenous knowledge became dissociated. This affected the way in which these collections have been interpreted, displayed and researched. The agency of Country was lost and the history of Aboriginal peoples' involvement in the making of collections forgotten.

In the last five to ten years there has been an emerging and well-deserved recognition of the agency of indigenous peoples across the world in collecting and documenting biocultural material (Schaffer et al. 2009; Newell 2010; Konishi, Nugent and Shellam 2015; Das and Lowe 2018). Specifically, this work has highlighted the existence of pre-colonial indigenous systems of land management, botanical and zoological knowledge, and the influence of this knowledge on the development of European scientific knowledge (Kimmerer 2013; Gammage 2021; Olsen and Russell 2019). However, a specific focus on *biocultural* collections in museums and their value and potential as material archives of indigenous knowledge has been until recently overlooked. James Oliver (2023, 9) puts forward relational knowing as a way of redressing the imbalances caused by colonial exploitation and advance, noting that the core of this work must be 'relevant, respectful and reciprocal'.

People and Plants and *Entangled Knowledges*, two research projects based on biocultural collections from Western Australia cared for by National Museums Scotland (NMS), have attempted to redress this imbalance. These projects have at their core relationship building and collaborative working and aim to be reciprocal and not extractive. They take as a starting point Linda Tuhiwai Smith's (1999) call to indigenise history and science through conducting a re-appraisal of material archives that enables a different kind of story to be revealed. The rest of this chapter will use these two projects as case studies for considering how this work can be undertaken for biocultural collections and what can emerge in the process. We use the term 'biocultural' in this chapter to mean collections that are plant, animal and cultural, that 'represent dynamic relationships among peoples, biota and environments' (Salick et al. 2014, 1).

Biocultural collections at National Museums Scotland

Founded in 1854, the Industrial Museum of Scotland (1855–64) was the first national museum to be formed outside London. The bespoke building known today as the National Museum of Scotland, on Chambers Street, opened in 1866 as the Edinburgh Museum of Science and Art (1865–1904), and was later renamed the Royal Scottish Museum (1904–85). In 1985 the Royal Scottish Museum and the National Museum of Antiquities of Scotland (1858–1985) were formally merged to become National Museums of Scotland, renamed again in 2006 to National Museums Scotland (NMS). Today, NMS cares for the collections of a number of early Edinburgh-based institutions, including the University of Edinburgh Museum of Natural History (EMNH), the Royal Society of Edinburgh (RSE), the National Museum of Antiquities of Scotland (NMAS), the museum of the Society of Antiquaries of Scotland (c.1780, SAS) and the New College Museum (later Free Church College Museum). The collections from these institutions joined the Edinburgh Museum of Science and Art and its later iterations at staggered points between 1865 and 1985. As these collections entered the museum they were artificially divided into disciplinary divisions and were further divided as departments refined the way they understood their collections. Today, the museum has four distinct departments: Science and Technology, Natural Sciences, Scottish History and Archaeology, and Global Arts, Cultures and Design. When the Edinburgh Museum of Science and Art opened its doors, its collections were presented together with the purpose of telling joint stories (Figure 11.2).

As the disciplinary departments refined so too did the displays and today the collections on display at what is now National Museums Scotland are, with some exceptions, presented in 'departmental stacks' running horizontally across the museum space.

In their 2018 article 'Nature read in black and white' Subhadra Das and Miranda Lowe discussed the results of a 2013 survey commissioned by the Natural Sciences Collections Association into visitor perceptions of natural history museums in the UK. They argued that the results showed that 'in the case of natural history museums … covert racism exists in the gaps between the displays', noting that in these museum displays there is more often than not a lack of cultural context to the material on display (Das and Lowe 2018, 8). The same could be argued in reverse for ethnographic museums in the UK in that often, particularly for displays of First Australian material culture, the link between cultural objects and Country is missing for the visitor. In addition, collections of Australian

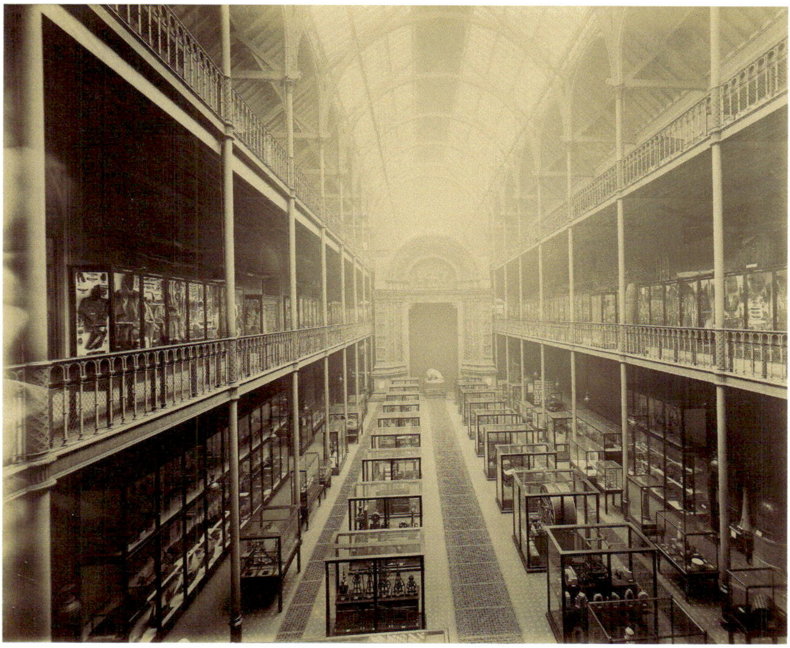

Figure 11.2 Interior of the Edinburgh Museum of Science and Art when it first opened. Source: National Museums Scotland, used by kind permission.

First Nations cultural material from across Australia are often brought together to represent 'Indigenous Australia', flattening the diversity of Country. When NMS reopened in 2011 after major capital works, the new permanent Living Lands gallery was opened. The aim of this new gallery was to explore the relationship between people and the land on which they live, considering how landscapes shape us and we shape them. The gallery, which displays cultural objects made by indigenous peoples from Canada, Australia, Tibet and Japan, includes some taxidermy specimens and ochre samples in an attempt to reassess the disciplinary divides that the institution had historically placed upon the collections. It remains ultimately, however, an ethnography gallery.

The 200 Treasures gallery, which opened at the Australian Museum in 2017, attempts to push this a little further and does a much better job of presenting holistic understandings of people and place through collections drawn from across the museum's holdings. Designed with the aim of showcasing '100 treasures of the Australian Museum' alongside the stories of 100 of Australia's most-influential people'[1] the gallery presents shields, cloaks and dance masks alongside shells, taxidermy specimens, animal skeletons and coral. On initially arriving in the gallery the visitor could be forgiven for thinking the Museum had recreated an 'Enlightenment Gallery', slipping back towards the universal museum. However, many of the displays on the ground floor of the Gallery focus on a holistic understanding of place. For example, a case on the Torres Strait Islands displays a turtle skeleton, a turtle carapace and masks that incorporate turtle shell in their construction to explain Country in the Torres Strait Islands.

In a paper delivered at the 2015 Museum Ethnographers Group conference, Cinthya Oliveira discussed how as scholarship moved towards the relatedness of things she was 'unsure about the relationship between nature and culture within the museum' (Livne 2016, 6), going forward. Conference host Inbal Livne suggested that the answer is perhaps 'to begin to re-forge the relationship between nature and culture outside the museum space' (Livne 2016, 6). The projects discussed below do just that, taking these collections outside of the museum, in an abstract sense, to reconnect them to people and place, to consider on one hand what role, if any, the collections have in the museum of the future and what other roles they might find outside it.

The *Entangled Knowledges* project

Funded by the Australian Research Council (2021–4) *Entangled Knowledges: Kaartdijin, science and history in the Robert Neill collection*[2] is a project that brings together a multidisciplinary team, including members of the Menang Noongar community, focused on a dispersed collection of fish and mammal specimens, art and material culture that was made by Robert Neill, a Scottish-born commissariat officer in the early 1840s, in Albany, Western Australia. The history of this collection and its distributed locations today reflect Robert Neill's family connections within the Edinburgh-based Wernerian Society and broader imperial networks in London. The Wernerian Natural History Society was formed in 1808 in Edinburgh and chaired by Robert Jameson. Naturalist Patrick Neill, Robert's father's cousin, was one of the founding members. Collections presented at the society meetings often became part of the collections of the University of Edinburgh's Museum of Natural History.

Today the collection is divided across departments in National Museums Scotland – the Natural Sciences Department holds 24 dried fish skins and two mammals, and the Department of Global Arts, Cultures and Design holds Menang material culture. In London, the Natural

History Museum (NHM) also holds seven dried fish skins, a mammal and a stunning portfolio of 69 watercolours of fish and reptiles. This collection has been framed in a Western science lens, concealing the rich Menangkaartdijin (knowledge) and stories that the collection also captures. The *Entangled Knowledges* project has been discovering ways to break down the disciplinary divides of 'nature' and 'culture' as well as understanding the culture of science in which it was framed in museums from the 1840s. The first aim of the project, however, has been to expose and return Menang knowledge to Menang families.

Between 12 May and 8 June 1841, Robert Neill received seven fish from Menang fishermen for a collection that he had begun several months earlier in the remote settlement of Albany, on the south coast of Western Australia. These fishermen – Wanuwar, Paddy, Moorianne, Toolegetwalu, Wallup, Munglewort and Marnett – speared fish in the shallows and rocky bays in the three harbours connected with their Country. They also caught fish in their stone fish traps, working with the tide, rocks, shrubby branches and deep fish knowledge to create a successful aquaculture. In a letter to his sister Ann in 1842, Neill acknowledged his reliance on the Menang men for the success of his collection: 'when hutted on the seashores', he wrote, the '[Aborigines] spear fish in large quantities ... Many of my curious [fish] were speared by the Aborigines which I could not obtain without them'.[3] Throughout 1841 Neill expanded his collection of fish to 59 specimens. He collected reptiles, mammals and botanical specimens as well. The fish collection was a large, communal one: in addition to the fish speared by the Menang men, Neill caught fish himself with a hook and line, soldiers and colonists brought him interesting fish hauled up in the town seine (a large fishing net), and sealers who lived and worked with Tasmanian and Victorian Aboriginal women across the Southern Ocean frontier brought him fish from their voyaging too.

The fish were preserved in a unique way, being stuffed with local sand to keep their shape. They were also preserved through Neill's artistic-scientific representations; each night, directly following the catch, Neill painted beautiful watercolours of the fish, for which the Menang were also present. The paintings represented the colour of the freshly caught fish before it faded and decomposed, but Neill was also careful to accurately depict the number of fins, scales and other important qualities needed for Linnaean identification. Neill collated the watercolours and connected them to the preserved fish skins with corresponding numbers from 1–59, in the order that they were caught. Neill wrote a descriptive account of each specimen which provides more detail from the Menang men about each fish, who killed it, if it was good eating, the fish habits and habitats and cultural and spiritual associations that the Menang had with them. Scales from the actual specimens were glued to both the descriptive account and onto the watercolour pages. The watercolours, descriptive account and a covering letter were sent to John Gray, Keeper of Zoology at the British Museum, sometime between 1841–3.[4] Today, this is collated and bound in a single portfolio held by the Natural History Museum, South Kensington.[5] The fish skins, however, were disconnected from the crucial Menangkaartdijinat at this point in time. We understand that they were sent directly to Edinburgh where they were given to the New College Museum (later Free Church College Museum).[6] In 1966 when the Free Church College Museum closed, the fish specimens were donated to the Royal Scottish Museum.

Today, 24 of the fish skins survive in the Natural Sciences collections at NMS. Despite this dispersal, the skins remain linked to the notes and paintings at the NHM in London by the Menang names on their labels. It was not just the physical collection of fish that the Menang helped Neill with, but the taxonomy of the collection. Neill reflected on this process in a letter to John Gray at the British Museum:

> In naming the fish, I have merely attempted to give the aboriginal and popular names known to the sealers and settlers. In obtaining the former, no little difficulty has been experienced. The younger natives generally giving different names to those of the elder; but finding the fish named by the latter more descriptive, I have, of course, in most instances, adopted them.[7]

This layering of Menang knowledge is evident on the sketches themselves. For example, Figure 11.3 shows Menang names recorded, the first most likely to be the name given to Neill by younger men – Kijetuck, and the second name 'Bebil', he records as 'since ascertained', suggesting that this is the name given by the older men, and the one he believed to be more accurate. He wrote that older men were, 'more minute in *species*; the younger often call very different fish by the same name'.[8]

The communal nature of this collection is reflected in the distinct knowledges represented. Directly on the sketch paper, Neill recorded Menang names for the fish, and the sealer and settler names given to the fish. Three separate cultural taxonomies. Later, ichthyologist John Richardson would add Linnaean names – a fourth taxonomy. They each tell a different story. What makes this 'natural history' collection so exciting is the strength and persistence of Menang knowledge throughout – on

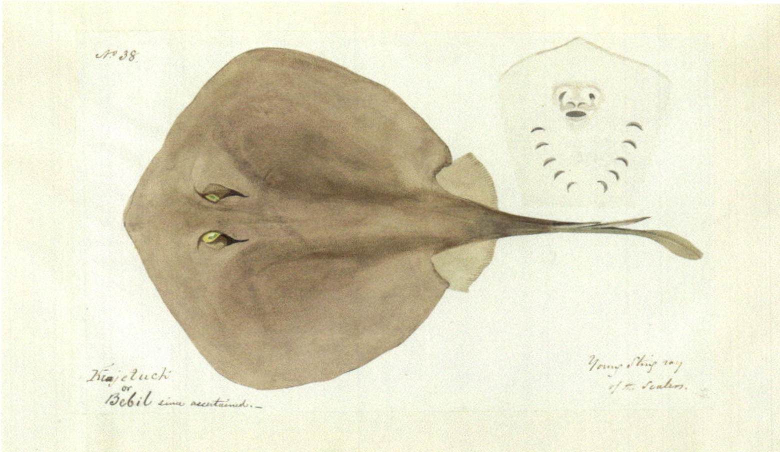

Figure 11.3 No. 38 sketch of Kijetuck or Bebil in Neill's portfolio.
Source: Library and Archives, Natural History Museum, London.

fish labels, sketches, archive notes and museum registers, Menang names were recorded by Neill. In addition to this focus on Menang knowledge, other aspects of the collection reveal a close relationship between Neill and the Menang, such as the Menang cultural objects that were most likely gifted to Neill by the same group of fishermen and their families and today are held in the collections of the Department of Global Arts, Cultures and Design at NMS. These collections came to NMS via the University of Edinburgh's Museum of Natural History and until recently were disassociated from the fish cared for by the same institution. Workshopping the fish names with the Menang community, it has been exciting to understand the ways in which Menang fish taxonomy was often directly related to their material culture.

For example, the Menang name given to the fish Tabeduck (Figure 11.4) is a reference to the word taap (knife), as the spike on the head of this fish closely resembles the taap (Figure 11.5), two of which are in the NMS collections. The knife depicted is thought to have been presented to the Wernerian Natural History Society by Robert Neill in 1849 and becoming part of the collections of the University of Edinburgh Museum of Natural History.

In reassembling this collection to return the Menang knowledge it holds to Menang families, the project is developing a method for working in a cross-cultural and interdisciplinary team that keeps Menangkartdijin (knowledge) at its heart. We are calling this the Kaartdijin Model. The team includes Menang community members, fish

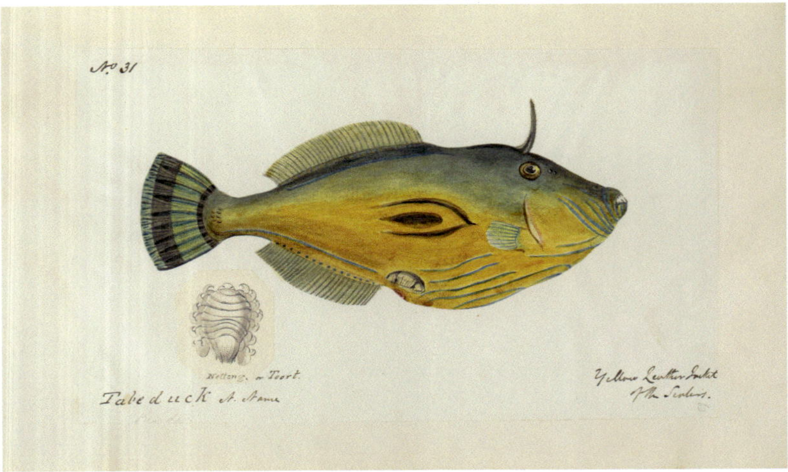

Figure 11.4 No. 31 sketch of the fish Tabeduck in Neill's portfolio. Source: Library and Archives, Natural History Museum, London.

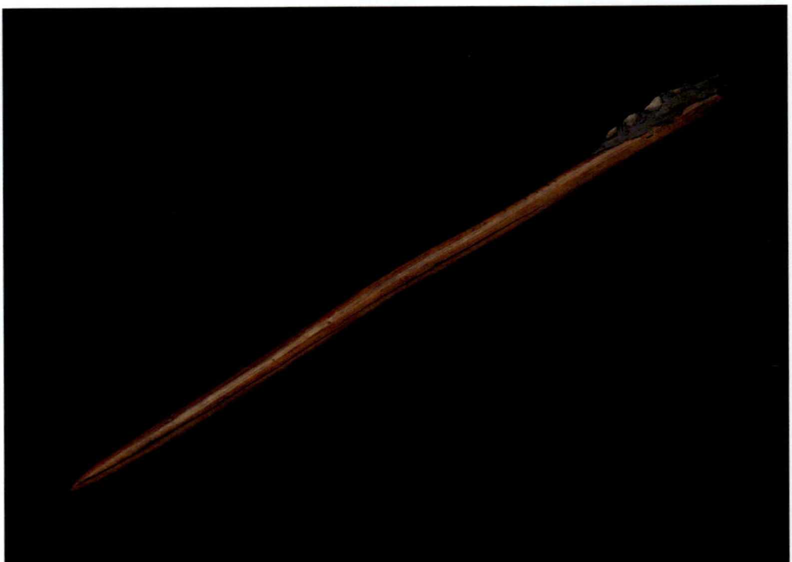

Figure 11.5 Taap (knife) in the collections of the University of Edinburgh Museum of Natural History. Source: National Museums Scotland, used by kind permission.

scientists, curators in anthropology and natural sciences, and historians. From the very beginning, the project has been governed and guided by the Albany Heritage Reference Group Aboriginal Corporation with CEO Lester Coyne at the centre. Since 2019 we have been unpacking the collection together, bringing our diverse expertise to bear on the cultures of history, science and knowledge that it represents. In April 2023, the team travelled to Edinburgh and London to get to know the collections and collection contexts at the NHM and NMS. We studied the beautiful watercolours of fish and reptiles in the NHM Archives and Special Collections, connecting fish names to sketches, uncovering links between the animals and objects. Sometimes our presence together in the collections encouraged a sharing of anecdotes or stories and other meanings were created. Some of the fish names weren't easily translated. Our team understands and respects that there is a diversity of meanings and knowledge across family groups. The fish called kojetuck reveals evidence of this (Figure 11.6).

This fish – its common name is snapper – was recorded by a later ethnographer, Daisy Bates. She recorded that the name kojetuck related to the Noongar tool *kodja*, or 'native axe'; the head of the fish, she wrote 'bears some resemblance to a koja'.[9] However, as one of the authors Shona Coyne explained during a workshop at NMS in 2023, *koitj* not *kodja* is the word for axe in Menang language, and *koitj* also refers to bones. The Menang men that named this fish for Neill also described

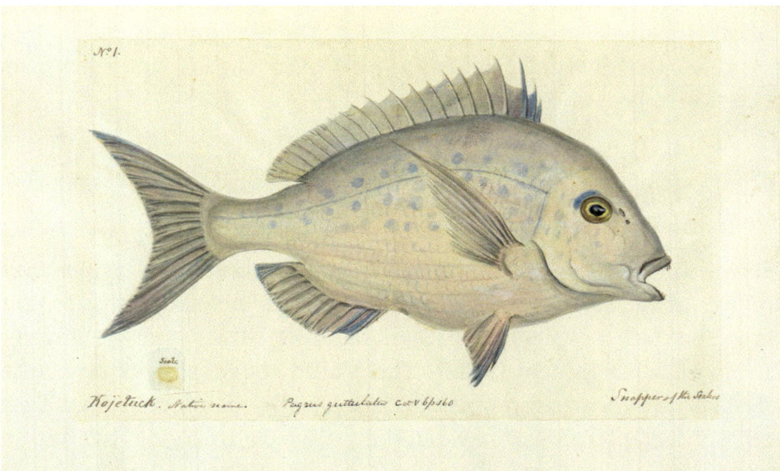

Figure 11.6 No. 1 sketch of the fish kojetuck in Neill's portfolio.
Source: Library and Archives, Natural History Museum, London.

this appellation, as Neill wrote: 'Kojetuck means fish with bones, which is very descriptive ... having very singular bones placed vertically in the neck connecting the dorsal spines to the back, resembling small tobacco pipes'.[10]

Clay tobacco pipes that were in use in the 1840s resemble the shape of a Menang koitj, and reveals Menang working cross-culturally, drawing on their own words and introduced objects to explain this fish and the meaning of its name to Neill.

Neill made very few Linnaean taxonomic observations himself, leaving that process to the scientists in the metropole, but his observations of Menang taxonomies and culture were entwined with his close friendships with Menang people, with whom he spent a lot of time. This is evident in one of the last additions to his collection. At a meeting of the Wernerian Natural History Society in 1848, Neill brought with him a live freshwater tortoise which he had caught in Albany. On the long voyage home to Edinburgh, Neill dunked it in the sea twice to keep it alive. At the meeting of the Wernerian Society, Neill referred to the tortoise as a Keelong, the Menang name for *Chilodina Longicollis*, the freshwater tortoise that still inhabit the area between Binnalup (Middleton Beach) and the wetlands in Albany. Neill told the scientists how he was taught to hunt for Keelong eggs by Menang. He had tasted the Keelong when sitting around Menang campfires, shown the way to cook and to eat it. This is a description of a close relationship in a colonial context. This particular Keelong from Menang Country survived in Edinburgh, living in the garden of Patrick Neill, until 1852, dying in the same year that Robert Neill himself died of yellow fever. The Keelong was stuffed and donated to the Museum in Scotland, but it has not been found among the collection today.[11]

The *People and Plants* project

People and Plants: Reactivating ethnobotanical collections as material archives of indigenous ecological knowledge was a collaborative project between NMS, Royal Botanic Gardens, Kew (RGB) and the Powell-Cotton Museum in Kent (PCM). The project sought to investigate the interplay between natural history (specifically, botanical collections) and ethnography collections building on a re-emerging interest in indigenous ecology and the value of ethnobotanical collections as material archives of indigenous ecological knowledge. The project aimed to create a conversation between academics, researchers, museum professionals, botanists and indigenous knowledge holders to establish the relevance of these collections in the present, and their potential applications for the future. The project, which was funded by a UKRI Arts and Humanities Council networking grant, held three workshops over the course of 2022 and 2023 at the three project institutions. Each workshop had a geographic theme which drew on the strength of the collections at the hosting institution: (1) Somalia at the PCM, (2) North-west Western Australia at NMS, and (3) Amazonian Brazil at RGB. The West Australian collection studied at NMS comprised of objects described as 'ethnobotanical' and 'ethnographic' within the museum records that were collected by Emile Clement from the Pilbara and the Kimberley regions of north-west Western Australia in 1895–1925.

Born in Prussia, Dr Emile Louis Bruno Clement (1844–1928) was a collector driven largely by the commercial opportunity of the collections he brought together. Originally working as a teacher at schools in England and Germany, he published a chemistry textbook, and excavated Bronze Age material at sites in Prussia which he sold to UK museums between 1877 and 1883. By 1895 he was working in Australia as a prospector, having established a mining lease at Toweranna, near the town of Roebourne in the Pilbara. Between mid-1895 and early 1900 he moved between north-west Western Australia and his family home in England, establishing other mining leases in the Pilbara. At the turn of the century, after his mining prospects started to decline in the Pilbara, Clement left and never returned to Australia (Coates 1999).

The majority of biocultural collections brought together by Clement come from the Pilbara and the Kimberley, and to a much lesser extent the Pacific Islands – often sourced from missionaries in the region. Today, the distributed Clement collections number approximately 1,600 and are comprised of cultural artefacts, shells, geology specimens and samples and botanical material. They are also accompanied by detailed letters, lists, drawings and photographs, and are spread out across institutions in the UK and Ireland, mainland Europe and the United States, including National Museums Scotland. In his doctoral study of Clements, Ian Coates (1999) identified two phases to Clement's collecting, the first between 1895 and 1900, when Clement was in Australia and collected the material both himself and with his son Adolf; the second between 1920 and 1928, when Clement sold collections from Western Australia and some from the Pacific Islands. These later collections were acquired through associates in the regions that the collections came from and mark a turning point for Clement, from 'entrepreneurial field collector' to England-based 'dealer' (Coates 1999, 124). Clement began selling to museums almost immediately he started collecting. While he was not trained as an anthropologist, he appears to have attempted to learn some of the First Australian languages in the areas he worked in and observed and recorded cultural knowledge. In writing a biography of Clement, Coates (1999) has shown that Clement's 'interest' in indigenous knowledge was driven by the demands of museums who purchased his collections rather than his own personal interest. In 1899, Clement published 'Vocabulary of the Gualluma tribe inhabiting the plains between the Yule and the Fortescue rivers'. In 1903 he then published *Ethnographical Notes on the Western-Australian Aborigines,* which included a map of eight Pilbara language groups from the speakers of which he had acquired collections, and was accompanied by a catalogue of ethnographic objects. The *Notes* were published in Leiden, where the Museum Volkenkunde had purchased objects from Clement in 1898. Clement's rudimentary understanding of the organisation of Western Australian Aboriginal social life was exposed by social anthropologist Alfred Radcliffe Brown, who criticised the confusion between skin groups and tribal names. Despite this, museums were keen to purchase collections from Clement and he understood how to sell them to museums. When contacting museums to offer items for sale he sent detailed letters and lists providing indigenous names for objects, as well as photographs showing First Australians using the collections. Sometimes he also sent detailed drawings of the objects for sale with descriptions of local use. He clearly understood that provenance was important to museums and aimed to provide as much detail to authenticate himself and these collections as possible. The collections he brought together are attempts to use material culture to characterise indigenous lifeways in the north-west of Western Australia at a particular moment in time, in a manner that demonstrates Country, so providing a 'package collection' for museums (Coates 1999, 133). On 30 September 1898, J. J. Buckley of the National Museum of Ireland noted that 'the collection of Australian ethno.[graphic] specimens sent on appro.[val] by Clement is very interesting: the uses of the different objects being given, as well as the native names, and, in most instances, the localities'.[12] As Coates has described, Clement's early mining activities and his 'practice of building up confidence and trust in his ventures was similar to the techniques he employed when corresponding with, and negotiating sales of objects to, curators' (Coates 1999, 120).

We do not know much about how Clement acquired the collections when he lived in Australia, especially given that there are only two surviving letters from him during his time there, but it appears likely that First Nations people were aware of his demand for material and that many of the items may have been provided in exchange. Clement was collecting from First Australian people who in their own lifetimes had lost access to their traditional country, having been invaded by settlers without their invitation or consent in the early 1860s. Across the new colonial frontier indigenous men, women and children provided labour to sheep pastoralists and pearlers – and were bound to the settlers by the Master and Servant Act (Paterson and Wilson 2009). Traditional food hunting was now a potential crime, as was unrestricted travel through country away from the newly defined sheep stations where indigenous people were bound to work.

Clement sold 185 objects from north-west Western Australia to the Edinburgh Museum of Science and Art and later the Royal Scottish Museum. The collections were sold in four lots, first in 1896, next in 1898, and then later in 1923 and 1925. It was the 1898 sale that specifically separated out botanical material from ethnographic, with 42 plant specimens sold, ranging from plants eaten, to spinifex used for baskets to wood used for shields. The 1898 sale was divided in the NMS register, with the first 74 objects described as an 'Ethnological collection from North-West Australia' despite it including 'spinifex gum', 'cullardie gum', red and yellow ochres and black pigments. The remaining 42 objects were described as 'Collection of Economic Botany from North-West Australia, comprising of native foods, woods used in making weapons and herbarium specimens', and from the listing provided in the register we can assume the collection mirrored the collections Clement sold to the Economic Botany collection at the Royal Botanic Gardens, Kew (RBG), and many of the plants listed in the NMS register are still to be found in the RBG's current collection.

Today, 154 objects remain in the collections of National Museums Scotland, 153 sold directly from Clement to the Edinburgh Museum of Science and Art and later the Royal Scottish Museum, and one further object, a pearl shell ornament which was transferred from the Wellcome Historical Medical Museum (WHMM) to the Royal Scottish Museum in 1953. Clement had sold material directly to the WHMM. Clement already had a relationship with the Edinburgh Museum of Science and Art, having sold European archaeological material to the Museum in 1887. His archaeological sales went to the national museums in Dublin, Edinburgh and Cardiff as well as to the British Museum and Ashmolean Museum, creating a reputation and a network for himself. It was these museums that he initially approached when he had First Australian collections to sell in 1896 and 1898, using these networks to further his reach across the UK and Europe. When the initial sales of Clement material arrived at the Edinburgh Museum of Science and Art they would have been held in its Art and Industrial Section. In 1901 this department split to form the Technological Department, and the Art and Ethnographical Department with the collections split accordingly. From the mid-nineteenth to the early-twentieth century, the Edinburgh Museum of Science and Art purchased, among other things, many collections comprised of ethnography and botany. Tayce Phillipson (2019) has noted that from 1860 there was an increase in botanical specimens collected by the museum when Thomas Archer, a botanist, became Director. Botanical specimens were collected both as economic botany and ethnobotany. Economic botany – the useful application of plants for industry – was of great interest to Empire; it focused on the global use of plants found across the world. Ethnobotany tended to have a more local focus, describing the relationships between people and plants, their symbolic as well as practical value.

This constant shuffling of collections and division into Western taxonomies meant that ultimately the economic botany collections and the ethnobotany collections, despite being intertwined, were divided between different departments. In 1928 it had been decided that the museum would stop collecting or maintaining botanical specimens and then, in 1938, there was a large-scale transfer out of collections labelled 'botany' to the Royal Botanic Garden Edinburgh (RBGE). However, in the late 1950s these were all disposed of by the RBGE, as in a botanical context they were perceived as having no scientific value. However, not all of the botanical material had been sent to RBGE, and some remains at NMS today – it was these remaining collections that formed the impetus for the *People and Plants* project.

In October 2022, funded by an additional grant from the Art Fund, two of the authors (Clark and Paterson) travelled to Roebourne to work with representatives from the Ngarluma Yindjibarndi Foundation (NYFL) on mapping the NMS Ngarluma and Yindjibarndi collections back onto Country. Led by Clinton Walker, a Ngarluma/Yindjibarndi man who runs Ngurrangga Tours, a group of 15 elders and emerging elders, researchers, art centre workers and photographers drove in convoy to the sites that Clement had located on his maps and written about in his notes.

The first site we visited was Inthanoona, a source of water and food and a known rock art and hunting site. During the colonial period sheep were also kept here. Ngarluma custodian Kerry Churnside showed us the

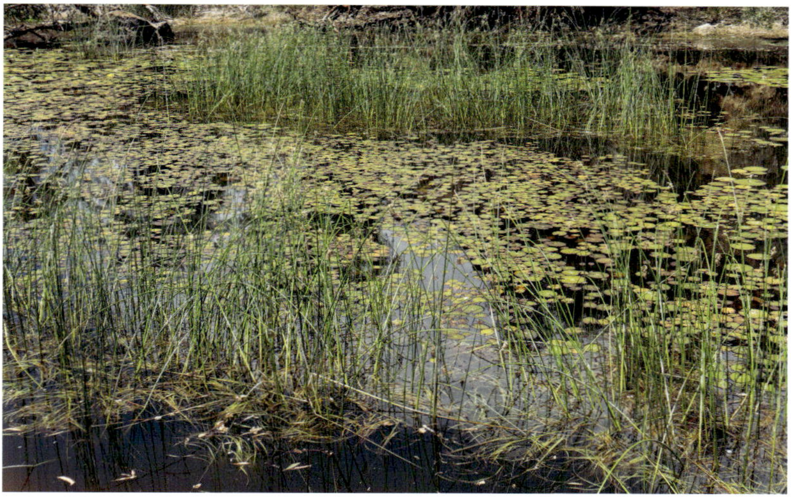

Figure 11.7 Inthanoona, Pilbara, Western Australia.
Source: © Alison Clark, 2022.

Garlun Marduwari (bullrush) which grows in the water at Inthanoona (Figure 11.7) and together we were able to match it to a box of dried bullrush bulbs, currently cared for by NMS, noting their use as a food source and part of a cultural artefact (Figure 11.8).

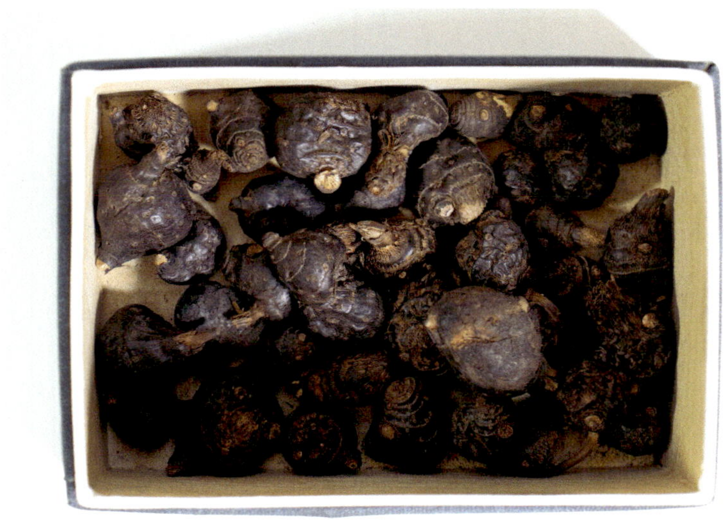

Figure 11.8 Bulbs sold by Emile Clement to the Edinburgh Museum of Science and Art in 1896. Source: National Museums Scotland, used by kind permission.

The bulbs at the museum had previously been misidentified as another plant, and now, through Kerry, we could situate these orphaned objects back in place and time. We were also able to match a photograph sold by Clement to Glasgow University to this site. The same day we also visited Balla Balla River, a fishing site, a historic site for ochre collection and a Thalu (increase) site for the white cockatoo. Tusk shells or scaphopoda, a type of mollusc, were collected from this area and then used to make shell necklaces, many of which were sold by Clement to museums, including NMS. This place would have been a site of activity and Clement may have come here specifically to collect these objects, and possibly to collect ochre as well.

Indeed, one of the materials collected in significant amounts by Clement from the Pilbara was ochre. He collected samples of red, yellow and white ochre pigment, storing them in small glass jars and selling them as sets to many museums, including NMS (Figure 11.9). Beyond the museum these collections also have huge potential for future collaborative research with the community.

During the visit to the Pilbara in October 2022 custodians expressed an interest in finding the sites where Clement had been taking ochre

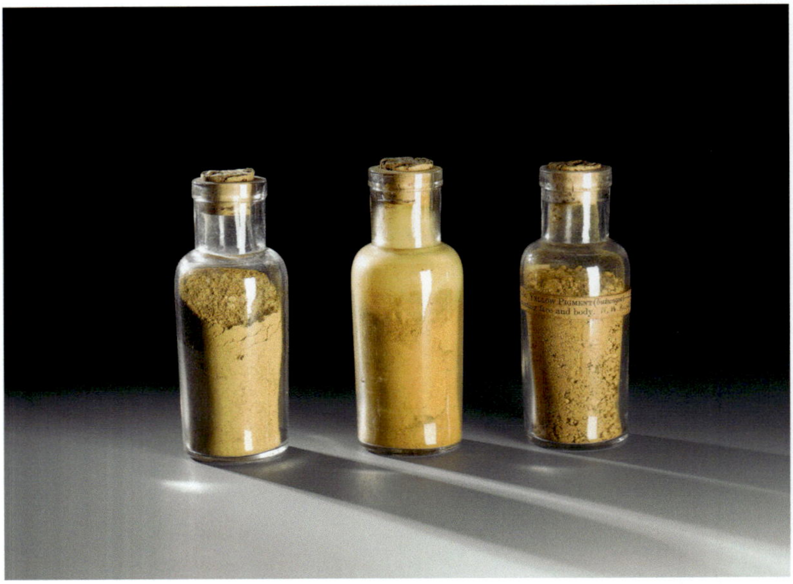

Figure 11.9 Ochre samples sold by Emile Clement to the Edinburgh Museum of Science and Art in 1898. Source: National Museums Scotland, used by kind permission.

samples from and whether research could help identify these sources. When Clark and Paterson visited Balla Balla, they gained permission to take ochre samples from the Thalu site – a place where ceremonies are conducted to increase the benefits of the natural world. Back at NMS the historic ochre samples collected by Clement have also been sampled, in order to compare samples. Both samples will be undergoing Scanning Electron Microscopy (SEM) analysis to characterise the ochre, which we hope will reveal geographic markers allowing for a comparison of contemporary and historic ochre sites on Country. This is a significant data set, as it may better reveal part of the vast networks of exchange through which materials such as ochre were circulated by Aboriginal people. This information could then be used by custodians such as Clinton as part of his tourism work and inform how and where ochre is collected for use in ceremony.

Clark (2013) has written about how when recording oral histories about cultural heritage, different places and spaces can produce different responses to the same or similar cultural heritage. The museum can be perceived as a sterile environment that alienates collections from their cultural contexts and in some cases holds an association with a repressive colonial power. The importance of situating these collections back into the landscapes they came from links then to the interconnectedness of people, land and culture. Objects regain their ancestral connections, and through these landscapes and objects people today can feel an emotional connection to this cultural activity.

Where next for biocultural collections?

Within museums, the need to categorise and define collections is currently driven by practical concerns of cataloguing, storage and care. How can the holistic nature of biocultural collections be incorporated back into the museum, and what needs to change in order for that to occur? Both projects discussed here have shown how bringing these dispersed collections back together and re-indigenising them has enabled a more informed understanding of how they function as part of a wider cultural framework. Working across disciplines, cultures and museums has revealed relationships that were previously misunderstood in the museum context. The on-Country Menang workshops held as part of the *Entangled Knowledges* project are helping NMS to understand the connections between their collection of fish, mammals and material culture that are currently separated in different departments. Working together across disciplines enables a richer, deeper understanding that could not be achieved within current discipline boundaries.

In the last five years there have been numerous people-centred cataloguing projects within museums in the UK, as well as outside of them,[13] with the projects collaborating with communities to build stronger connections between communities and collections. Projects based outside of the museum act as a kind of online data aggregation, with the priority being to understand collections from within indigenous frameworks rather than affecting change in the museum. Projects based within the museum however, such as those at the Powell-Cotton Museum and the Pitt Rivers Museum (Lawther 2023; O'Brien Backhouse 2024) looked outwards at wider society and non-Western ways of knowing while reflecting inwards to change museum practice (O'Brien Backhouse 2024), Both *People and Plants* and *Entangled Knowledges* looked to explore who museum databases were for and whether their museum documentation practices were fit for purpose when considering how they represent communities. They were about collaborating with communities on the minutiae of museum work: cataloguing, care and interpretation, all crucial for affecting change within museums. Learning from the methodology and practice of people-centred cataloguing projects is one way to address biocultural collections in museums, and the research that comes out of projects like *People and Plants* and *Entangled Knowledges* can ensure that indigenous knowledge is embedded in how museums approach the collections going forward.[14]

Notes

1 *200 Treasures of the Australian Museum*, https://australian.museum/exhibition/200-treasures/. Accessed 23 January 2024.
2 https://www.nms.ac.uk/collections/departments/global-arts-cultures-design/projects/entangled-knowledges. Accessed 14 September 2024.
3 Excerpt from a letter from Robert Neill to his sister, Ann Neill, 29 October 1842, within a letter from Patrick Neill to Sir George Clerk, 13 June 1843. Australian Joint Copying Project M584, M985–M986. Collections held by the Scottish Record Office (as filmed by AJCP). Papers of Clerk Family of Penicuik, Midlothian (GD 18).
4 Neill, 1843.
5 Neill, Robert, 1845, '67 original watercolour drawings of mammals, reptiles and fish found at King George's Sound, Western Australia, and in its neighbourhood; accompanied by manuscript notes', Natural History Museum, South Kensington, Zoology Artwork, 88 f NEI.
6 *Fishes, Contents of New College Museum*, vol. 3, New College Museum. University of Edinburgh Centre for Research Collections, A.1.12.3.
7 Neill, 1845.
8 Neill, 1845.
9 See her 'Vocabulary from Albany & Denmark. MSS. Muning wonga Jäkbăm ♀ about 70 and now dead. BUMBLEFOOT about 75. also dead', http://staging.bates.org.au/text/40-260M.html. Accessed 14 September 2024.

10 Neill, R., letter to John Gray, Keeper of Zoology, British Museum, (no date), in Neill, '67 original watercolour drawings of mammals, reptiles and fish found at King George's Sound, Western Australia and in its neighbourhood; accompanied by manuscript notes', Natural History Museum, Zoology Artwork, 88 f NEI.

11 A note confirming the donation of the Keelong is in the University of Edinburgh Museum of Natural History daily books, 18–19 May 1852, 44–5, stored at National Museums Scotland Library.

12 J. J. Buckley to Mr White, Minute Paper, 549/AT, Science and Arts Institutions, Departmental Letter File, 1898, Vol. 2, National Museum of Ireland Archives, Dublin, 'Arts and Industry box, 180'.

13 For example, *Digital Benin* (2020, https://digitalbenin.org/) and *Recollecting Rapa Nui* (2020, https://www.indigen.eu/projects/core-projects/recollecting-rapa-nui), both following on from the long-running Reciprocal Research Network (https://www.rrncommunity.org/). All accessed 14 September 2024.

14 The research for this chapter was funded by the Arts and Humanities Research Council UK, Australian Research Council and the Art Fund Jonathan Ruffer Curatorial Grant. The authors would like to thank colleagues who have commented on earlier drafts of this work or informed its thinking: Mark Nesbitt, Inbal Livne, Kevin Guiness, Kerry Churnside, Kate Oosterhof, Andrew Kitchener, Glenn Moore, Ross Chadwick, Lester Coyne and Larry Blight.

References

Alberti, Samuel. 2012. *Nature and Culture: Objects, disciplines and the Manchester Museum*. Manchester: Manchester University Press.

British Association for the Advancement of Science. 1874. *Notes and Queries on Anthropology*. London: E. Stanford.

Clark, Alison. 2013. 'Conversations in Country: Tiwi and Yirandali Indigenous Australian collections at the British Museum'. PhD thesis, King's College London.

Clement, Emile. 1899. 'Vocabulary of the Gualluma tribe inhabiting the plains between the Yule and the Fortescue Rivers'. *Journal of the Royal Anthropological Institute of Great Britain and Ireland* 29: 192–6.

Clement, Emile. 1903. *Ethnographical Notes on the Western-Australian Aborigines*. Publications of the Royal Ethnographical Museum at Leiden, Series II, No. 6. Accessed 15 September 2024. https://nla.gov.au/nla.obj-17270043/view.

Coates, Ian. 1999. 'Lists and Letters: An analysis of some exchanges between British Museums, collectors and Australian Aborigines (1895–1910)'. PhD thesis, Australian National University.

Das, Subhadra and Miranda Lowe. 2018. 'Nature read in black and white: Decolonial approaches to interpreting natural history collections'. *Journal of Natural Sciences Collections* 6: 4–14.

Freshfield, D. 1889. *Hints to Travellers*. London: Royal Geographic Society.

Gammage, Bill. 2021. *The Biggest Estate on Earth: How Aborigines made Australia*. Sydney: Allen and Unwin.

Haraway, Donna. 1989. *Primate Visions: Gender, race and nature in the world of modern science*. London: Routledge.

Healy, Jacqueline. 2018. *The Art of Healing: Australian indigenous bush medicine*. Melbourne: University of Melbourne Press.

Herschel, John Frederick William. 1849. *A Manual of Scientific Enquiry: Prepared for the use of Her Majesty's Navy and adapted for travellers in general*. London: John Murray.

Kimmerer, Robin Wall. 2013. *Braiding Sweetgrass: Indigenous wisdom, scientific knowledge and the teachings of plants*. Penguin: London.

Konishi, Shino, Maria Nugent and Tiffany Shellam. 2015. *Indigenous Intermediaries: New perspectives on exploration archives*. Canberra: ANU Press.

Lawther, Kathleen. 2023. 'People-centred cataloguing'. Accessed 31 January 2024. http://www.kathleenlawther.co.uk/wp-content/uploads/2023/01/people-centred-cataloguing-final.pdf.

Livne, Inbal. 2016. 'Nature and culture in museums: An introduction'. *Journal of Museum Ethnography* 29: 3–10.

Martin, B. 2023. 'Thulu – our ancestor'. In *Shipping Roots: Plant journeys through empire*, edited by Keg De Souza, 32–5. Edinburgh: Royal Botanic Garden Edinburgh.

Newell, Jennifer, 2010. *Trading Nature: Tahitians, Europeans and ecological exchange*. Honolulu: University of Hawaii Press.

O'Brien Backhouse, Meghan. 2024. 'Who do you say you are?'. *Journal of Museum Ethnography* 37: 34–51.

Oliver, J. 2023. 'Archipelagic seeds: Ceding power is seeding possibility'. In *Shipping Roots: Plant journeys through empire*, edited by Keg De Souza, 7–10. Edinburgh: Royal Botanic Garden Edinburgh.

Olsen, Penny and Lynnette Russell. 2019. *Australia's First Naturalists: Indigenous peoples' contribution to early zoology*. Canberra: NLA Publishing.

Paterson, A. and A. Wilson. 2009. 'Indigenous perceptions of contact at Inthanoona, Northwest Western Australia'. *Archaeology in Oceania* 44: 98–110.

Phillipson, Tayce. 2019. 'Collections development in hindsight: A numerical analysis of the Science and Technology collections of National Museums Scotland since 1855'. *Science Museum Group Journal*. Autumn. https://dx.doi.org/10.15180/191205/001.

Salick, Jan, Katie Konchar and Mark Nesbitt. 2014. *Curating Biocultural Collections: A handbook*. London: Kew Publishing.

Schaffer, S., L. Roberts, K. Raj and J. Delbourgo. 2009. *The Brokered World: Go-betweens and global intelligence, 1770–1820*. London: Science History Publications.

Smith, Linda Tuhiwai. 1999. *Decolonizing Methodologies: Research and indigenous peoples*. London: Zed Books.

Stocking, George W. 1992. *The Ethnographer's Magic and Other Essays in the History of Anthropology*. Madison: University of Wisconsin Press.

Thomas, Nicholas. 1991. *Entangled Objects: Exchange, material culture and colonialism in the Pacific*. Cambridge, MA: Harvard University Press.

Thomas, Nicholas. 1994. *Colonialism's Culture: Anthropology, travel and government*. Princeton: Princeton University Press.

12

Digital heritage technologies and issues of community and cultural restitution in 'new style' ethnographic museums: a digital update

Michael Rowlands and Graeme Were

Over a decade has passed since the pivotal RIME[1] conference that took place in Rome 2013, at which a network of museum curators, anthropologists and other scholars from ethnographic museums across Europe and North America came together to discuss the legacies of ethnography museums and their political, social and cultural futures. A focal point for discussion and conference proceedings was the question: 'Do ethnography museums need ethnography?' – a provocation intended to raise debate about the continued relevance of anthropological methods in the research, display and interpretation of ethnographic collections in museums. Such a question was particularly poignant given that the Musée du Quai Branly in Paris, one of the most recently opened ethnography museums in Western Europe, had abandoned anthropology in favour of an 'arts premiers' ('first nations' arts') approach and the subsequent aestheticisation of ethnographic objects in its permanent galleries (Price 2007).

These institutional shifts in reframing ethnographic collections reflect a broader crisis in ethnographic museums around their colonial legacies and, equally, their futures. As witnesses to looting, theft and dispossession, Hicks (2020) sees ethnographic museums as sites of 'extreme violence' and advocates for the wholesale return of collections to communities of origin. Thomas (2021), meanwhile, resists this singular viewpoint by instead envisaging them as sites for both good and harm, much like temples or other religious buildings.

While the intensification of the 'culture wars' has heightened attitudes towards the decolonisation of ethnographic museums and 'arm's length' approaches to the state governance of museums, what emerges in the wake of these polarising debates is a call for a new form of cosmopolitanism, bringing a new form of interaction, learning and engagement with cultural heritage. This call for cosmopolitanism is one that supersedes earlier ideas of universalism that were established to counter calls for the return and restitution of cultural property. Indeed, in a recently published work that addresses the notion of cosmopolitanism in ethnography museums, Kuper (2023) introduces the idea of a cosmopolitan museum as one that is rooted in the values of expert knowledge, evoking the sentiments raised at the RIME conference. As a kind of new cosmopolitanism, he proposes that ethnographic museums around the world should act as sites for the dissemination of cultural knowledge and understanding of other peoples and their ways of life – echoing Cuno's (2008) earlier work on the primacy of museum objects as sources of human creativity and ingenuity.

Kuper's approach (2023) is heavily nuanced by the legacies of nineteenth-century European epistemologies, when it was assumed that objects were not merely to be looked at but were sources of meaning and knowledge not available to the untrained observer. But if this formed an earlier legitimation for the purpose of the museum to display as many objects as possible, then critiques of such 'elitist' views of the museum object led to the dramatic reduction of objects on display in museums through the latter part of the twentieth century leading some, like Conn (2010) to ask, 'Do museums still need objects?'.

This question brings us back to the driving force of our original contribution to the RIME conference and the key topic we intend to address in this chapter: that is, how do digital technologies re-frame these debates about the contemporary status of museum objects and their relation to meaning and knowledge? Our call for a deeper questioning from within museum anthropology occurs in the context of renewed questioning of the pedagogy involved in what some see as the consequent trivialisation of ethnographic displays (as evident in the Musée du Quai Branly), but it also occurs in museological debates about the status of the physical object as a synecdoche for larger bodies of knowledge (rather than an entity in itself that can be appreciated aesthetically).

This renewed questioning converges with the experience of the ethnographic museum in the two non-European contexts of Melanesia and West Africa. We suggest that digital technologies offer an alternative rendering to the issue of object status and knowledge, chiefly because there is ambiguity over the relation between the real object and the digital reproduction. In societies such as Melanesia and West Africa the attitude to physical objects is often, to put it mildly, deeply ambivalent, if not frankly fearful, if an encounter with artefacts (from the past) involves touching or coming into close proximity with them. The introduction of digital technologies, we argue, in these contexts may not be seen as poor substitutes for the authenticity of the 'real thing' but, rather, raise another set of questions about their digital coding and their potential re-entry into community life. In the two cases we deal with, the digital recording, while capturing an object or song in time and space, may be perceived in a distant and less threatening manner if it cannot be touched or sensed in the same way as real objects or, for a song, as it is articulated in ritual performance. If digital recordings are perceived in such a distant way, and appearing less real, does this distancing allow these new digital cultural forms to be internalised and hence facilitate their re-making, thus engineering their return in community life? What then, is the implication of this for understandings of physical objects and for that matter, the ethnographic museum, in an era where the very definition of the museum and its purpose is under scrutiny?

Digital imaging, museum objects and cultural safeguarding

A considerable body of work has demonstrated the application of digital heritage technologies to cultural preservation such as the use of digital imaging to restore paintings, sculptures and monuments that have suffered degradation or are at risk of destruction (Kalay et al. 2008; Macdonald 2006; Stanco et al. 2017). These studies have largely focused on contexts in which value is placed on heritage objects themselves – monuments, artworks, film, and artefacts – and not on the preservation of skills and knowledge required to reproduce them. Conversely, when safeguarding has become the key framework from which to approach intangible heritage (Akagawa and Smith 2019; Alivizatou 2012; Stefano and Davis 2017), only a handful of scholars have investigated its growth alongside digital heritage as a tool for education, particularly from the perspective of transmission, inheritance and intergenerational learning (Alivizatou 2019; Alivizatou-Barakou et al. 2017).

Yet we also know that the status of museum objects is located not primarily in the finished object as something to be admired or appreciated, but in the skills and practices required to reproduce them. As Bolton (2001) has demonstrated in relation to one of the purposes of the Vanuatu Cultural Centre in Melanesia, the collections play a necessary role in cultural revitalisation. The capacity to provide ni-Vanuatu women with access so they can engage with collections in order to learn how to reproduce them – unpicking, touching and manipulation being the tactile skills alongside visualisation and internalisation of techniques and material processes – gives the institution and its collections value and significance. Similarly, in Queensland, Australia, many digitisation projects have focused on developing repositories for documenting and safeguarding Aboriginal knowledge, set up through the State Library and its networks across the state, to safeguard social histories for future generations. What is at stake, or so it appears, is the sustaining of memory and the potential to re-craft the past through documentation of skills, practices and local knowledge that underpins their value for local communities.

The rise of digital imaging technologies calls into question the status of museum objects, not just in the Walter Benjamin sense of debates about aura and mechanical reproduction but by bringing into focus the potential for reproduction, through access to and possession of knowledge of how to reproduce an object housed in a museum. Brown (2008), for instance, points to the potent nature of these technologies and the ethics of reproduction. Drawing on her work with Māori communities in New Zealand, she describes how certain types of ritual protocols are required for the care and curatorship of Māori *taonga* that have been reproduced through 3D imaging and printing. In the same way physical objects in the Vanuatu Cultural Centre offer opportunities for learning, possession and revitalisation. Access to the original provides Māori people with the opportunity for reproduction, so creating a facsimile that is potent and alive, and can be used to revitalise cultural heritage.

Scholars have focused on emancipatory and democratising principles of digital technologies in the context of museums and archives through a capacity to enable new forms of access, empowerment, self-representation and visibility (Christen 2006; Hennessy et al. 2013; Lambert 2002; Srinivasan 2012; Were 2013). But this chapter questions how this is played out on the ground in a Melanesian society which is regarded as image-based; one where images and rights to produce images are valued more highly than the actual physical objects themselves (Kuechler 1987). In Francophone Africa, the status of digital objects is affected by the colonial history of the attribution of objects as the product of *ethnies* or tribes (cf. Ravenhill 1996). Formal discontinuities between object types were seen as the product of a natural order of ethnic groups with their own distinct language, history and material culture. Today, the future of museums created under colonial rule in Francophone West Africa remains heavily influenced by the legacies of this tribal idea of the ethnographic object, and by a postcolonial antipathy, promoting displays of national unity through festivals and performance relating to the precolonial past (Arnoldi 2006).

Here, we discuss two digital heritage projects – in New Ireland, Papua New Guinea and in Mali, West Africa – to illustrate how the digital economy, in this more nuanced and re-contextualising state, is transforming the way communities access and internalise their cultural heritage held in ethnographic museums around the world. These two projects raise further issues about the application of digital technologies to transmissions of intangible heritage, and the potential of digital technologies for the restitution of cultural knowledge. In relation to debates on the future of the ethnographic museum, it raises critical questions about the future rendering of the museum and the kinds of values attached to virtual spaces as a particular kind of 'safe' and secure medium for the archiving of cultural heritage.

The Mobile Museum project, New Ireland, Papua New Guinea

The Mobile Museum was established in January 2012 as a pilot project to facilitate remote access to Queensland collections of historical ethnographic artefacts for Nalik people living in New Ireland, Papua New Guinea. It was initiated after Naliks in the community expressed a strong desire for access to museum collections for cultural revitalisation purposes. There had been several attempts to build a museum in the community but these had not materialised, probably because the type of objects that people had anticipated placing inside the museum were considered 'dangerous heritage' by the Nalik (Kingston 2007). People in the Nalik community asked for support in assisting them to find ways to deliver remote access to museum collections from their homes. They believed digital heritage technologies could be used to revitalise traditional practices (*kastom*) which, in turn, would strengthen Nalik society. As a vehicle for transmitting *kastom*, Naliks believed that digital technologies would reconnect young people to their cultural heritage (as they were already familiar with mobile phones and so forth) and act as an antidote to social problems in the community. Developing a digital platform would enable them to observe and learn about culturally relevant artefacts housed in museums in Australia and elsewhere.

As a pilot project, the Mobile Museum project provided access to 10 digital objects for people living in the Nalik community, with the future aim to expand this number by training people in the community to scan their own objects. Few, if any, Nalik people visit museum collections, either in the Papua New Guinea (PNG) National Museum and Art Gallery in Port Moresby or further afield in Australia. This is because of the costs involved, rather than for lack of interest. What made the project so timely was the introduction of mobile telecommunications in New Ireland. Up until around 2005, the only phones available on the island were a few landlines in the provincial towns of Kavieng and Namatanai. This changed dramatically with the introduction of mobile telecommunication masts by Digicel, a Jamaican-based Caribbean telecommunications company. This infrastructure resulted in fledgling mobile and internet coverage along the coastal villages, although the reception could best be described as patchy.

A significant aspect of the project was how the development of digital infrastructure on the island coincided with community concerns to adapt the technology for cultural safeguarding purposes. Within the Nalik community, cultural preservation and revivalism are important issues as *kastom* was perceived to have declined over the last century (Were 2010). In particular, a focal point of debate has been the sustaining of the *malangan* tradition, for which northern New Ireland is famous. *Malangan* carvings objectify the material histories of the Nalik people, and they play a central role in funerary rites. The carvings make visible relations to land and to history through the incorporation of clan totem designs (Kuechler 1987). However, unlike those carvings which are placed on display in glass cases for public consumption in Western museums, *malangan* carvings are not on show or on display in everyday life; they are significant because they are displayed fleetingly during events leading up to the culmination of mortuary feasting in communities. Clan members inspect the carvings during these ceremonies, ensuring that the image is witnessed and remembered, before the carvings are removed and destroyed, either by burning on a fire or through decay when left in a special place in the forest. Using 3D digital images of actual *malangan* carvings, the Mobile Museum project thus offered a chance for Naliks to reconnect with museum collections, through witnessing and observation which would otherwise be inaccessible to them. It also provided them with the opportunity to access historical collections dating to the late nineteenth century which exhibited diverse styles of carving and to compare them with objects that are currently made in Nalik society, which were seen as less sophisticated. Crucially, the project opened the door for Naliks to learn about past carving practices through visual analysis of museum collections in three-dimensions, so creating a new way in which people could learn about their past through a digital archive rather than rely on memory alone.

Across the Pacific, specialised craft learning is a relationship of respect, taking place between master and apprentice through direct observation and practice (Borofsky 1987). In the Nalik community, young men learn how to become a carver (*aitek*) under the guidance of a master carver in the community. Apprentices take many years of study and practice before they attain the credentials to practice carving and receive commissions from land-holding clans. Carvers produce the *malangan* figures in ritual seclusion, out of sight from the public, after undergoing a short period of fasting. The carver builds a special leaf shelter inside which he sets to work on a seasoned piece of softwood, *azabaf*, performing a series of ritual offerings during the production process. The image of the carving is said to appear to the carver in a dream. Each *malangan* is carved using a template design, so they are recognisable as a particular class of carving that includes clan totems and designs which are known to the carver and knowledgeable people in the community. Those carvers attaining renown in New Ireland are known as master carvers, which generally infers that their work is also known by art dealers in the West.

Carving is not without risks: as Gell (1998) has argued in his analysis of Marquesan art of the eastern Pacific, carving ancestral figures introduces an element of danger to the carving process, since each design involves invoking ancestral images which are considered potent. This is exemplified in the New Ireland case, as the softwood from which *malangan* are carved is said to be inhabited by a spirit. This spirit can, it is said, devour humans, and so Naliks treat *malangan* with utmost caution and respect. Many stories circulate in New Ireland society about carvers who had disrespected the *malangan* spirit and died as a consequence. Thus, one of the important questions of the project was to consider whether digital images mediated ancestral power, and if so, how would this be managed?

Such were the community concerns of the risk of loss that their initial plan was to give young people in the community liberal access to the software platform, so that they could freely learn about *kastom* and *malangan*. The idea was that the software could allow young people to learn about technical and material knowledge, such as the types of leaves and pigments used in adorning carvings, and to view this in full colour 3D. The technology itself was important as the community felt that, given young people's interest in computers and mobile phones, the technology provided a means to re-engage them and to redress the perceived dissipation of traditional knowledge.

Scanned as 3D digital objects, the images of *malangan* could be physically manipulated using computer and mouse, appearing as lifelike textured images of spatial complexity, with the high levels of surface detail necessary to conjure the notion of replica or simulacra (Baudrillard 1994). Digital objects are distinguished from 2D images or photographs by the fact that they evoke the sensorial capacity of depth perception, which imbues images with a spatial dimension and tangibility that lets one 'feel around' objects. Naliks expressed this in being able to 'see' the object in its entirety rather than rely on photographic formats which offer only partial views. As Adam Kaminiel, a *malangan* carver described it, 'The software allows me to see the carvings as light in my eyes', suggesting that the image can be possessed and known in its totality. Indeed, in Kuechler's analysis of *malangan* (Kuechler 1999), she has argued that it is not just the totemic emblems and symbols on the *malangan* that are significant, but also the topological surfaces; the interweaving openwork spaces in-between that reveal to Naliks the memory of relations to land. Thus, the benefit of the 3D digital objects resides in their potential to be understood spatially and in their totality, enabling them to be remade.

Community design and digital objects

Several scholars have demonstrated how digital tools need to take into account traditional structures in cultural documentation projects so as to be 'respectful' to local sensitivities and values (Christen 2011; Verran and Christie 2007). Our participatory methodology involved working closely with diverse groups within the Nalik community to decide how digital access should be mediated and structured, working collaboratively as a form of 'appropriate museology' (Kreps 2008). Guided by this approach, the project team ran a series of consultation workshops – called 'awareness' by Naliks – in several villages along the east coast Nalik area, during 2012 and 2013. The awareness campaign involved visiting several communities and meeting with community leaders, senior men and women, women's groups, teachers and school children to ascertain the ways in which the software could be designed to benefit local people and meet their needs. Some meetings were held with several people, and some on a one-to-one basis. Many of the villages along the east coast of the Nalik area were consulted several times over the duration of the project, and key reference groups from each village were established to help implement the project. In all cases, Nalik people used the application and were asked to comment on its functionality, how they could use it (if at all), and potential implementation issues. Problems of implementation that arose included the number of computers in the community, who had access to the software and questions about digital literacy.

The workshops allowed the group to compile a set of specifications for the design of the software. First, local carvers requested 'hot-spots' to be placed on certain artefacts. These were small blue spots placed on the surface of carvings on points of special significance to Naliks. On clicking the hot-spot, a detailed photograph could be launched in the digital platform. This allowed the carvers to see details of the carving and tool marks. Second, some participants requested that they be able to compare two 3D objects side-by-side. This would allow them to analyse variation in design and style of artefacts. Third, Naliks were interested to know the size of the artefacts – so a matchbox was inserted into the virtual environment to ascertain scale. Fourth, Naliks also wanted museum documentation to be included in the digital platform: this allowed users to learn about the object's collecting history and contextual information. Finally, a notes field was introduced on the request of Naliks so that annotations could be added to the 3D images.

In addition to these design requirements, the workshops also tested the technical limitations of the project. The main constraint was the lack of bandwidth on the island, which meant that large file sizes were slow to transmit, even though the project used relatively small file sizes (5–10 MB). Furthermore, while most Nalik people owned mobile phones, less than 10 per cent of these phones had the capacity to access the internet. Even within this small number of owners, the cost of purchasing internet data was relatively quite high, which was a deterrent. As a result, it was decided to provide access to the digital platform, in the first instance, using a CD-ROM (rather than a website providing access to the 3D images) which would be distributed widely in the Nalik communities. This meant that people could study the *malangan* without internet access, and without the burden of having to find money to buy internet data to download the images.

A key stage in the development of the project involved a 10-day visit to Brisbane by the two Nalik partners in the project Martin Kombeng and Adam Kaminiel, in late 2012. The aim of their visit was to access the New Ireland collections in the museum storage at the Queensland Museum and the Anthropology Museum at the University of Queensland, the institutional project partners. Together, both institutions housed in total around 200 artefacts of New Ireland provenance. The two Nalik men were able to view all the objects and spent time identifying which of the artefacts came from the Nalik area. They were able to recognise various designs, styles and motifs, and explained the types of observational skills required to identify the design of each object. They also analysed the different types of materials used in constructing the objects, some of which had dried or decayed, making them hard to recognise. They then selected 10 objects for 3D scanning. Nine of these were *malangan* carvings and the other was a fish trap. The two men chose the carvings because they were ritually important and recognisable (from totemic symbols and motifs) from the Nalik area; while they argued that the fish trap was typical of New Ireland as a region.

Returning the digital objects

In 2013, the software was distributed as a CD-ROM by a group of senior Nalik men to key community members such as chiefly men (*maimai*), senior men and women, and schoolteachers in villages along the east coast of northern New Ireland. The distribution was described in terms of the Nalik term *poxai* which means 'obligation' or 'reciprocity'. The thinking behind this form of distribution was that Nalik people could seek Martin or Adam for help with any unforeseen technical problems related to the software. As Martin Kombeng exclaimed at the time, 'People will come to Adam for help with the software; they know where to find him. They know he is knowledgeable about computers!'.

The return of *malangan* as 3D digital objects generated much discussion about how the technology could be used to revitalise *kastom*. One of the main responses by the community was to note how contemporary carvings produced in communities appeared less sophisticated in technique and style than those held in museum collections, dating from the nineteenth and early-twentieth centuries. Some of the *malangan* styles, such as a clamshell *malangan*, seemed unfamiliar to many people. For many Naliks, there was an obvious reluctance to admit to having lost knowledge of their own *malangan*, because of its strong link to social history and land, which are both crucial resources in New Ireland. This dynamic became manifest during a workshop in one village along the east coast when a crowd of about 20 men and women gathered around the laptop computer to eagerly see the *malangan*. As the crowd dissipated after about 30 minutes, a senior man came forward to say how people were so pleased to see the *malangan*. He then explained how he thought that each person was interested in spotting their own clan's *malangan*. Once they discovered that their own clan's *malangan* was not captured in the software they were, he claimed, no longer interested. Certain styles of *malangan* are owned by land-holding clans and rights to produce them are closely policed. Much like mortuary ceremonies, where *malangan* carvings are placed on display, the software provided the opportunity to study and internalise the images publicly. But rather than do this as a form of ephemeral art, the software introduced a more permanent state for observation and analysis, much in the way *malangan* today are no longer destroyed or left to rot in the forest. Their digital 'permanence', in effect, became another reminder of the threat of their loss – while at the same time, their abstraction from mortuary feasting meant that they were presented as distant and less threatening (much in the way we describe in the next section on Mali, West Africa).

These events give an indication of the way in which knowledge and control over ancestral images are articulated in Nalik society, as a form of management control (Harrison 1995). On the one hand, clans risk the problem of losing their *malangan* if the carvings are not reproduced from memory. Because there are fewer carvers and fewer new carvings made in contemporary society, specialist knowledge of carvings is at risk of dissipation or loss. On the other hand, attempts to liberalise access to ritual images through digital technologies are met with sanction or trepidation, as I now describe.

The software interface developed as part of the Mobile Museum project promised a future return of *kastom* and the possibility of cultural completeness (Rowlands 2002). Among the Naliks, the technology drew an emotional response. Community members used the expressions *bringimbek* (returning) or *putimbek* (putting back) to describe how the software application was providing access to museum collections many thought had been lost to time. Many people said that the community were (at the time) unprepared for the physical return of *malangan*, chiefly because the carvings held in ethnographic museums were too powerful for people to handle today. The digital renderings of *malangan* were seen as lifeless entities, but entities that held the potential to be re-made into objects that were potent and dangerous.

While the project set out to widen access to ancestral images, the distribution of the software instead reinforced existing ritual policies and practices. Technology (rather than the images) became the object, with CD-ROMs hidden away as a resource for future use and action (Isaac 2008). In some cases, the distributed CD-ROMs were used by individuals, rather than collaboratively, almost mimicking the model of traditional learning – with the computer taking on the traditional role of master (Borofsky 1987). Some Naliks described how they would study the digital objects alone on their laptops, sometimes using pen and paper to sketch out designs for themselves, in a kind of mentorship relationship to the software.

Attempts by senior Naliks to take control of the technology then took another turn. In 2013, the senior men showcased the Mobile Museum project to the Kavieng District Council and received funding to formalise the integration of a Council of Chiefs (*maimai*) within the Tikana region (Tigak, Kara and Nalik-speaking areas). Their discussions were the culmination of a series of meetings held over the previous five years, in which the provincial government had workshopped the idea of establishing a formal organisation to administer a traditional leadership system in the region. The government was alarmed about the declining moral and social fabric of New Ireland society and believed that traditional learning would instil core values.

The senior men's success in acquiring funding and recognition was based on their claim that because these images were 'out there' and therefore required some management, the *maimais* had the traditional authority to take control of them. They argued that the proposed Council of Chiefs needed the finances and resources to administer the digital objects with the purchase of computers and mobile phones. In effect, by showcasing the successes of the Mobile Museum, the *maimais* were able to secure further access to state funds and equipment.

This serves as a reminder, given the emergence of digital technologies and digital assets within an image-based economy, of how the virtual world becomes an important resource from which to make claims and contestations. The archive, in its virtual guise, is a potent space containing digital objects, and offers opportunities for the re-making of museum objects that now reside in ethnographic museums far away. For Naliks, the value attached to these digital objects, at least in the present, resides in their capacity to be worked with and interrogated visually, safely and securely, and at arm's length, in contrast to the historical collections held inside museum storage facilities which harbour the spirits of ancestors.

The Digi-Dogon project, Mali, West Africa

Our second case study examines Digi-Dogon, a project established to assemble an accurate record of both material and immaterial Dogon heritage. We neither assume the completeness of such a record nor a critical conclusion on what constitutes Dogon heritage. Our aim has been more limited, to show that we can develop a common approach that will integrate ethnographic studies of both oral traditions and material culture with the databases created for museum collections of Dogon material culture. A methodological issue concerns the fact that Dogon oral and visual documentation has come mainly through ethnographic fieldwork, while the databases of Dogon material culture derive from studies of museum collections. There is no reason to assume some harmonious merging of these different data sets. As seen, we noticed the antipathy between the role of the museum in being set up to serve the role of French colonial administration and an ambivalent status for the contribution of ethnographic fieldwork starting, in the case of the Dogon, with the recordings of Griaule and Leiris and the Dakar-Djibouti expedition of 1931.

While these remain deeply problematic questions, those raised by our work concern more how we can contextualise the oral, visual and material sources of knowledges of Dogon practices. The group in question are the Dogon of Mali, whose larger environment, the area of the Bandiagara Escarpment, has been recognised as a UNESCO World Heritage site since 1989, and whose material culture has acquired a great renown. The immaterial heritage on which this project focuses, concerns the *bajani* song cycle that is part of the traditional Dogon funeral and forms a pivotal feature of the ritual complex. The song texts claim a respected lineage, stemming as they do from a nineteenth-century Dogon prophet and blind poet, Abirè, with a huge reputation in the area. The song cycle forms the core of the *yuyana*, which is the first stage of a complex of Dogon funerary rituals. This is followed several years later by the *dama*, most associated with mask performances and, finally and only after 60 years, by the *sigi*, as the final confirmation of ancestral status.

The Dogon *yuyana* funeral consists of a spectacular series of rituals lasting five days and four nights, engaging the whole of a village plus quite a few of its neighbours. During the second night of the funeral proceedings the *bajani* are sung, from 11 p.m. until 6 a.m. the next morning – no less than seven hours of solid singing. The majority of this long string of related songs is attributed to Abirè, who roamed the plains and cliff area of the Dogon, probably in the second half of the nineteenth century, and left a huge legacy of funeral chants and prophecies. The texts usually pertain to mourning and loss, interspersed with political commentaries of the times, reflections on life in general and on death in particular, while also referring to prophecies. Within the songs there are two subtypes, the *sèmbeleni* and the *bajana*. The first are the older songs, already in existence before the appearance of the prophet, and are usually more straightforward songs of mourning. Singing the *sèmbeleni* fills the first part of the night and sets the stage for the emergence of the *bajana* (meaning 'large *baja*'). The latter form the major part of the songs, and are of a more complicated nature, with parabolas, symbolic speech and oblique allusions to folktales and prophecies. The song performance of the *bajani* inside the *yuyana* contrasts clearly with the *dama*, which is replete with masks, and during which very few songs are performed in any case. The *bajani* is never sung during a *dama*. The same also holds for the *sigi*, which is quite different again from the funeral proceedings – and also features no masks. The *sigi* can in fact be seen as a second birth, and is never mentioned in the *bajani* texts – in which death is central.

The *bajani* as a whole forms a mandatory item in each collective funeral and its songs are of the general African solo+refrain-by-choir type. All couplets share the same melody line, thus binding the various texts together in one melodious whole. The singers are not professionals but men from the village who happen to like singing, and have a 'good ear', meaning a good memory and a voice that carries in the night. Participating in the various funerals they learn by doing, singing first the refrains and later moving on to performing the solo parts. As *bajani* singers they are well respected in the village, as no funeral can be performed properly without them, and people appreciate that the singers keep the legacy of Abirè alive. Abirè was also a 'Homer of the cliff', giving prophecies in the villages he visited that predicted what would happen on that particular spot in the future; these prophecies form a second oral source for this pivotal figure.

Due to Christianisation and Islamisation, the traditional funerals are disappearing and the Dogon themselves realise that this cultural heritage is under threat. The present wave of Islamist troubles in Mali increases the pressure on any traditional performance and thus speeds up the disappearance of this heritage; it in fact forms a huge and immediate threat for the whole Dogon culture, and surely for all rituals, including the *bajani*. In the plains, where Abirè came from, the performance as such has almost become a thing of the past, since Islamisation has made deeper inroads already. Increasingly the cliff-side villages, where the tradition is still alive, are following suit.

In the absence of a political body, who can speak of cultural preservation for 'the Dogon'? In the traditional setting, no one could – but that has changed. The core of the Dogon area was recognised in 1989 by UNESCO as a World Cultural Heritage site due to its 'exceptional combinations of natural and cultural elements'. But, of course, it is the nation-state of Mali that has claimed and is the recipient of this status, so to a considerable extent Mali 'owns' this heritage. As Dogon country is part of Mali, their cultural wealth is seen as Malian property, though with the explicit recognition that the local Dogon people are the actual repositories of the heritage.

In the early 1990s however, an ethnic association for the Dogon appeared on the scene: Ginna Dogon, the 'large Dogon house', with the aim of 'protecting and promoting Dogon culture'. The association defines its goal in Dogon terms as *tèm* customs, meaning to preserve and as well as to promote what the Dogon themselves see as their core customs. *Tèm* means 'found': that is, applying to all items of behaviour and all objects handed over by the previous generation. The core activity of the association is the organisation of triennial cultural festivals. These Dogon *semaines culturelles* are held every three years in one of the larger towns in the Dogon area, or more recently in Bamako for safety reasons, and on the whole are quite a success. Thus, for the Digi-Dogon project, Ginna Dogon is a natural partner alongside the state of Mali and the local communities. The association has been closely involved already in the first documentation phase that has resulted in the essential document on Abirè and in the formulation of the first publications, and is keen to stay involved. Consultation on the form and accuracy of the digital archive has had to filter through a process of authorisation by the National Museum of Mali, Ginna Dogon and leaders in local Dogon communities.

For the Digi-Dogon project the aim was to digitise, store and make accessible the songs on the website of Universitaire Bibliotheken Leiden and a website formed by the National Museum of Mali in Bamako, in combination with the publications of the texts and the commentaries of three books: one French and two English.

The issue in both digitisation and heritage definition is whether one can lift a piece out of a puzzle and preserve that piece in isolation. With material objects such is always the case, and with intangible heritage it is hard to avoid. The Malian Ministry of Culture works to define the *sigi* as the whole of Dogon ritual in order to apply for UNESCO Intangible World heritage. In this ritual, which is in effect an orchestrated walk through each village, it would be hard to lift out one specific item. Also, the preference for the *sigi* is that it occurs very rarely, once every 60 years; the next cycle – lasting five years – is due from 2026 onwards. The structure of the *yuyana* is different, since it consists of a series of more-or-less related rituals, among which features the *bajani*. Here the central elements could well be preserved or nominated individually. Many imagistic religions have rituals that have this kind of 'Lego' character, in which pieces can be taken out and replaced by other similar ones.

Obviously, the total setting of the ritual complex is important, but an exaggerated holism prevents any total preservation and would result in unmanageable digital renderings. Digitisation increases the out-of-context character of the songs, so in preserving them there is an obligation as part of the digitised version to furnish as much as possible the cultural and ritual context. Mask dances are an example that comes to mind, like the Gelede or Ijele masks in Nigeria, which are now recognised as world heritage. What is sampled in this case are the headpieces – rather than the costumes, song texts and dancing footage. What is stressed instead are the organisations behind them, and the social arenas that these mask dances address. The loose structure of the *bajani* singers is an important aspect; there is no secret society in Dogon (despite some reports mentioning them) and this is a conglomerate of choice, with the singers defining themselves as the adopted sons of the poet/prophet. This renders the continuation of the heritage fragile, but also opens other venues for popularising the songs.

A special problem for intangible inheritance is that what we record and digitise is the outward form of a performance: the sound, the text and some photos or film. As for any performance, this is just the shell. The important thing is that the performance has to be done – as with any ritual – and that participation/presence in it is crucial: as singers, drummers – or even as a sleeping female audience. The *bajani* singers face two tests: to sing, and to stay awake. This is a challenge that is hard to reproduce outside the ritual situation. Description and recording do not equal existential participation, and meaning inevitably shifts when one concentrates on the observable, digitised form. As with the ritual context, this existential aspect can only be touched tangentially in words. Any performance is dynamic, and any rendering of the *bajani* songs is new, in that it will differ from all other performances. It has to be recognisable – the prototype Gell (1998) mentions – but it is not identical to an earlier version, and neither should it be. Inevitably, but also desirably, the contingencies of the performance have to be integrated into the selection of songs, the variations on the texts and, especially, the many detours a lead singer makes to comment on bystanders and fellow singers – like, for instance, the wake-up song which is part of the night. The singers are very much aware of this, and when they reproduced in the analytic sessions some songs to highlight meaning and variation, they stressed that theirs would be 'different'.

The challenge for any heritage recognition is that cultural productions are produced on the spot, not reproduced. Variation and change, often in the way of cultural drift, are standard, and should be included. What this means for digitisation is clear: digitisation should end in an invitation to produce more digitised versions of the heritage, and versions that are adapted to the present. But the reality is somewhat different. The core recordings were made by Van Beek in one village, Tirreli, from 2005 to a final long moment in 2016. A selection of singers was brought to Bamako in 2018 (because of the security situation) for 'analytic sessions'. For a month, singers explored the recordings, making comments and further renditions to be added to the corpus. It is these digital recordings that will be preserved. What will be added over time are further recordings made in different villages by researchers who would be able to make them as part of returning to celebrate the rituals in their own place. But we have to realise that what we are doing is materialising a record of funerary songs that are intended to be fleeting and to some extent imaginary. The singers themselves will assert that each of their renditions is different from ones before and from others. Yet we recognise certain harmonies. It is establishing what these may be that become the object of publishing texts on them which of course then fixes them in a literary form (Van Beek et al. 2022).

Our problem of a fleeting creativity constrained by a search for consistency is a little different from any others arising from the recordings of oral traditions, songs and performance. The fact that it is a problem is of course a product of heritage definition, cultural ownership and authorisation. In a wider West African setting, the songs of griots praising their patrons were traditionally produced for and owned by a patron. UNESCO's recognition of griots as embodiments of intangible cultural heritage (ICH) and the requirement of copyright laws in Mali, has meant the songs now belong to griots, rather than their patrons who personalise the contents (cf. Röschenthaler and Diawara 2016). The legalities of copyright serve to legitimise and objectify both the contents of songs and their ownership. But in the case of the *bajani* singers, their creativity is fleeting and of the moment. It is the audience, now in particular women but also the elders of lineages, who express their satisfaction in the performance. Once done, it is completed, and the efficacy of the performance is judged in terms of the likely satisfactory transmission of the spirit of the dead person to a future stage of potential ancestorhood. It is this focus on the value of performance and acts of making (that once done are not just forgotten but, quite literally, are required to be recognised as completed in order for their efficacy to be established) that can be recognised quite widely in Africa as a principle of oral potency.

The digital does therefore contribute to the impact of heritage and cultural copyright in objectifying performance as a single discrete event. In a way this continues the role of the French colonial museum in Francophone Africa in collecting, decontextualising and exhibiting ethnic objects. But the difference is not so absolute. Very few objects are tied into a performance of the *bajani* – in fact, just two: the round calabash drum and the lances of the dancers. The round drum, *barubɔ*, is only used by the Dogon inside the *bajani*, and nowhere else. But the drum as such is not of Dogon origin, since it probably stems from their neighbours. This type of spherical hand drum is known elsewhere in Mali, Burkina Faso and Ivory Coast. In Bambara it sometimes accompanies the *balafon* and is called *bara* or *bendre*. It is also known among the Senufo of North Ivory Coast as a part of *bolonye* ensembles performing in the suburban neighbourhood of Korhogo, the main city in the Senufo area. These drums are more common in the region of Dianra, Ivory Coast, which has a mixed population of Manding and Senufo. Yet both in Bambara and Senufo musical culture, such drums are rather marginal, and do not form the core of their musical repertoire; the same holds for the Dogon, who only use these drums for the *bajani*.

Historically, the appearance of the *barubɔ* on the Dogon scene might well coincide with the life of Abirè, and the *bajani* text repeatedly refers to the fact that in such and such a village which the prophet visited, the *barubɔ* had been introduced already. So, in this calabash drum we have a material heritage that is nowhere central, but occupies a similar niche in several cultures, that is, that of a symbol for a specific subfield of creative action. The other material item of interest is the lance, *saru*, which is not restricted at all to the *bajani* performance but is used in the funeral dances in general. In the *bajani* dance, which is just a gentle anticlockwise shuffle around the heap of stones at the centre of the public grounds, it is central: one should have something like a lance in hand, and the texts repeatedly define the singers as bearers of the lance. During most of the night and in many song texts, the spear is the symbol of being alive: anyone dancing with his spear, or who leans on his lance, is alive, and – just as important – awake. Avoiding sleep is crucial during the night, so the spear is the symbol of being wide-awake and alive. These objects derive their symbolic valence from their use, they must be there and yet they are not assigned any special potency or sacredness for the Dogon.

Digitisation is for the Dogon a matter of exaggerating paradoxes already implicit in the performances of ritual. The idea of a single digital version of the *bajani* songs, including their tones and accuracy, would be inconsistent with the plurality of performances consistent with the commemoration of the person, their village and collective setting. Yet there must be a set of recognisable *bajani* songs and the role of certain objects that follow the story of the blind prophet Abirè and his journey at one time through the whole of the Dogon region inventing and disseminating new songs. The singers see no inconsistencies, since whichever performance is recorded is unique and therefore, to an extent, consistent with the pride they take in their creativity.

What, may we ask, are the implications of these inconsistencies for the remaking of ethnographic museums in Francophone West Africa? The colonial legacy of the collection and display of objects for identification and administration of ethnic groups is matched by antipathy to experiencing fixed items in display rather than what has already emerged as the preference for festivals and performance as a better way of catching the physical presence and harmonies of collective life. Ambivalence towards objects that may be chosen for display because of the ritual or aesthetic qualities is not only a matter of antipathy to the fixing of images. Both the images and the making of them evoke a danger of who might see them or learn from making them. The learning of funerary songs, done at night, is to sing with the singers at a funeral, and the repetitions form the essence of memory and their transmission. Their content and allusions concern death, and the bush or wilderness as a source of power; also, the bush is seen as where death resides and from where it comes into the village. Managing the power of sounds, images and objects is widely attested in many African settings (as is seen already in Melanesia) and their display is a subject of fear and ambivalence (Rowlands 2011; Mew 2012). This does not mean that their sources will necessarily be hidden or excluded from use. There are no secret societies, for instance, among the Dogon and objects like masks used in rituals like the *dama* will be seen as useful for a particular purpose and then simply disposed of. As Van Beek et al. (2022) show, the Dogon archive is a collective memory based on repetition and doing rather than any attention to recording and documentation for the creation of an archive. On the contrary, the museums established by the French and maintained by the nation-state in Mali through the postcolonial time retain a different power-inflected mode as an archive. Mbembe describes the colonial archive as possessing an inescapable materiality; the status of its architecture is entangled with the documents and objects housed within (Mbembe 2002).

How one reconciles the reality of Dogon social performance, its recording in digital form and the housing of it in a national museum in Mali and in Leiden is on the one hand a question of access, and on the other one of protection and preservation. The question of access is, as we have seen, part of the wider solution of how a virtual world becomes part of an image-based economy and gains added value to that derived from the incident of its first performance. Embedded in the museum, do questions of authorisation, copyright and ownership deprive the performance from the singers or the community origins of the *bajani*, *dama* and *sigi* rituals? Optimistically, we could anticipate that the ethnographic museum in Francophone West Africa becomes the repository of re-articulated values of material and social knowledge that would be accessible online, as in our Dogon case (and also in New Ireland, Papua New Guinea), to children in education and their wider communities.

Conclusion

To conclude the chapter, our two case studies underscore the ambiguity of digital heritage and the tensions that emerge in providing access to cultural forms that are either re-inserted back into community life – through digital return and reuse – or held in its archival form as a distant and abstracted audio recording or 3D rendition for future generations to learn, make and perform. As we demonstrate, fixing images and sounds through digital recording and archiving introduces the paradox of permanence and stasis. This paradox thrives on a fixed notion of object as reference point in time rather than a processual notion of creativity and innovation that emerges through its re-imagining and re-articulation in ritual and performance. Answers to our paradox provide accounts of how very different ideas of creativity and materiality exist to shape discussions of authenticity in heritage practices on a global scale. In the Melanesian cases, authenticity in the image does not constitute a fixed material form; rather, an idea evoking flux and change in relation to claims of ancestry and belonging. Here, the digital image liberates the local from some universal idea of their material culture as fixed in time and space. On the other hand, the knowledge that collections in museums can be drawn upon as a permanent repository (not only for inspiration but also for creating certain buffering constraints on what might pass as a suitable image) provides a security in the safe hands of an archive defined in a quite material, architectural sense. Paradox equally erupts in the fleeting creativity of Dogon funerary chants and ritual performance. In the recordings available we have a momentary fixity in what constitutes their content, and yet – given the longevity of memory involved and the emphasis on lineage elders as the proper practitioners capable of articulating words in song – we can only assume that creativity is valued. But what constrains and maintains cohesion within a Dogon sense of a complete performance is the reference to an event, and in particular the travels of Abirè, a prophet whose words are articulated and responded to in any given performance. The role of the round drum and holding the spear in dancing around the appropriate setting of a pile of stones in each Dogon village, highlights the potential for a shared value of archive and museum.

As our cases from Melanesia and West Africa emphasise, the notion of transmission that digital technologies enables is framed within a liberal idea of access and emancipation from colonial collections, as part of a decolonising process that values dialogue and giving back alongside the responsibility to safeguard and protect.

Seen in this light, the new cosmopolitan museum, in its digital and highly mobile guise, may become a beacon for hope in the restitution debate, and for communities seeking to reinvigorate their own cultural heritage and identities; yet, at the same time, these digital interventions run the risk of raising issues of rights over reproduction and authenticity. As we argued at the RIME 2013 conference, what appears important for the future of the ethnographic museum is the recognition that what is important is not necessarily the matter of the real objects being returned to communities, after all. This is because by unmooring images and sounds from their material being in former colonial institutions, their digital guise gives access to future generations who can potentially reproduce them. But if the cosmopolitan museum is going to repatriate objects and other cultural expressions in their digital form, then there needs to be a better understanding about how the virtual world gains value and potency once utilised or deployed in community life (as political interventions), and how cultures of display that recontextualise objects and sound recordings render them impotent or ambivalent.

As we enter an era where decolonisation and restitution are at the forefront of ethnographic museum agendas, characterised by closer collaborative partnerships and the gradual return of looted artefacts, the notion of the cosmopolitan museum is a reminder of how museums still cling on to ideological visions of the object rooted in terms of its originality and authenticity, and as a source of knowledge and understanding which somehow needs to be understood and unlocked. Quite what this means for social performance and creativity in communities where these digital assets emerge through digital intervention projects, such as the two described in this chapter, has implications for collective memory and the mapping of time into which these objects emerge alongside the patterns of community life. This means that these new digital objects could begin to take on a more real, authentic and significant presence than originally anticipated, as prototypical images that fix ritual and performance into a pattern of the authentic past rather than the collective memory of the present.

Note

1 Réseau International de Musées d'Ethnographie/International Network of Ethnographic Museums (RIME).

References

Akagawa, N. and L. Smith. (eds). 2019. *Safeguarding Intangible Heritage: Policies and practices*. London: Routledge.

Alivizatou, M. 2012. *Intangible Heritage and the Museum: New perspectives on cultural preservation*. Walnut Creek: Left Coast Press.

Alivizatou, M. 2019. 'Digital intangible heritage: Inventories, virtual learning and participation'. *Heritage & Society* 12 (2–3): 116–35.

Alivizatou-Barakou, Marilena, and 24 others 2017. 'Intangible cultural heritage and new technologies: Challenges and opportunities for cultural preservation and development'. In *Mixed Reality and Gamification for Cultural Heritage*, edited by M. Ioannides, N. Magnenat-Thalmann and G. Papagiannakis, 129–58. Cham: Springer. https://doi.org/10.1007/978-3-319-49607-8_5.

Arnoldi, M. J. 2006. 'Youth festivals and museums: The cultural politics of public memory in post-colonial Mali'. *Africa Today* 52 (4): 55–78.

Baudrillard, J. 1994. *Simulacra and Simulation*, translated by S. F. Glaser. Ann Arbor: University of Michigan Press.

Bolton, L. 2001. 'The object in view: Aborigines, Melanesians, and museums'. In *Emplaced Myth: Space, narrative, and knowledge in Aboriginal Australia and Papua New Guinea*, edited by Alan Rumsey and James Weiner, 215–32. Honolulu: University of Hawai'i Press.

Borofsky, R. 1987. *Making History: Pukapukan and anthropological constructions of knowledge*. Cambridge: Cambridge University Press.

Brown, D. 2008. '"Ko to ringa ki ngarakau a te Pakeha" – Virtual Taonga Maori and museums'. *Visual Resources* 24 (1): 59–75.

Christen, K. 2006. 'Ara Irititja: Protecting the past, accessing the future – indigenous memories in a digital age'. *Museum Anthropology* 29 (1): 56–60.

Christen, K. 2011. 'Opening archives: Respectful repatriation'. *The American Archivist* 74 (1): 185–210.

Conn, S. 2010. *Do Museums Still Need Objects?* Philadelphia: University of Pennsylvania.

Cuno, J. B. 2008. *Who Owns Antiquity? Museums and the battle over our ancient heritage*. Princeton: Princeton University Press.

Gell, A. 1998. *Art and Agency: An anthropological theory*. Oxford: Clarendon Press.

Harrison, S. 1995. 'Anthropological perspectives: On the management of knowledge'. *Anthropology Today* 11 (5): 10–14.

Hennessy, Kate, Natasha Lyons, Stephen Loring, Charles Arnold, Mervin Joe, Albert Elias and James Pokiak. 2013. 'The Inuvialuit Living History project: Digital return as the forging of relationships between institutions, people, and data'. *Museum Anthropology Review* 7 (1–2): 44–73.

Hicks, Dan. 2020. *The Brutish Museums: The Benin Bronzes, colonial violence and cultural restitution*. London: Pluto Press.

Isaac, G. 2008. 'Technology becomes the object'. *Journal of Material Culture* 13 (3): 287–310.

Kalay, Y., T. Kvan and J. Affleck. 2008. *New Media and Cultural Heritage*. London: Routledge.

Kingston, S. 2007. 'Dangerous heritage: Southern New Ireland, the museum and the display of the past'. In *The Future of Indigenous Museums: Perspectives from the Southwest Pacific*, edited by N. Stanley, 47–69. Oxford: Berghahn.
Kreps, C. 2008. 'Appropriate museology in theory and practice'. *Museum Management and Curatorship* 23 (1): 23–41.
Kuechler, S. 1987. 'Malangan: Art and memory in a Melanesian Society'. *Man* 22: 238–55.
Kuechler, S. 1999. 'Binding in the Pacific: Between loops and knots'. *Oceania* 69 (3): 145–56.
Kuper, A. 2023. *The Museum of Other People: From colonial acquisitions to cosmopolitan exhibitions*. London: Profile Books.
Lambert, J. 2002. *Digital Storytelling: Capturing lives, creating community*. Berkeley: Digital Diner Press.
Macdonald, L. 2006. *Digital Heritage: Applying digital imaging to cultural heritage*. London: Elsevier.
Mbembe, A. 2002. 'The power of the archive and its limits'. In *Refiguring the Archive*, edited by C. Hamilton, V. Harris, J. Taylor, M. Pickover, G. Reid and R. Saleh, 19–27. Dordrecht: Springer Netherlands.
Mew, S. 2012. 'Rethinking Heritage and Display in National Museums in Ghana and Mali'. PhD dissertation, SOAS University of London.
Price, S. 2007. *Paris Primitive: Jacques Chirac's museum on the Quai Branly*. Chicago: University of Chicago Press.
Ravenhill, P. L. 1996. 'The passive object and the tribal paradigm: Colonial museography in French West Africa'. In *African Material Culture*, edited by M. J. Arnoldi, C. Geary and K. Hardin, 265–82. Bloomington: Indiana University Press.
Röschenthaler, U. and Mamadou Diawara. 2016 (eds). *Copyright Africa: How intellectual property, media and markets transform immaterial culture goods*. Canon Pyon: Sean Kingston Publishing.
Rowlands, M. 2002. 'Heritage and cultural property'. In *The Material Culture Reader*, edited by V. Buchli, 105–14. Oxford: Berg.
Rowlands, M. 2011. 'Of substances, palaces, and museums: The visible and the invisible in the constitution of Cameroon'. *Journal of the Royal Anthropological Institute* 17: S23–38.
Srinivasan, R. 2012. 'Rethinking digital cultures and divides: The case for reflective media'. *The Information Society* 28: 24–36.
Stanco, F., S. Battiato and G. Gallo. 2017. *Digital Imaging for Cultural Heritage Preservation: Analysis, restoration*. Florida: CRC Press.
Stefano, Michelle and Peter Davis (eds). 2017. *The Routledge Companion to Intangible Cultural Heritage*. London: Routledge.
Thomas, Nicholas. 2021. 'Don't trash talk museums at this perilous time: We must adapt – not throw away – our cultural heritage'. *The Art Newspaper*, 2 June. Accessed 16 September 2024. https://www.theartnewspaper.com/2021/06/02/dont-trash-talk-museums-at-this-perilous-time-we-must-adaptnot-throw-awayour-cultural-heritage.
Van Beek, W. E. A., Oumarou S. Ongoiba and Atimè D. Saye. 2022. *Singing with the Dogon Prophet*. Lanham, MD: Lexington Books.
Verran, H. and M. Christie. 2007. 'Using/designing digital technologies of representation in Aboriginal Australian knowledge practices'. *Human Technology* 3 (2): 214–27.
Were, G. 2010. *Lines that Connect: Rethinking pattern and mind in the Pacific*. Honolulu: University of Hawai'i Press.
Were, G. 2013. 'Imaging digital lives'. *Journal of Material Culture* 18 (3): 213–22.

Index